Business Intelligence and Mobile Technology Research

Business Intelligence and Mobile Technology Research: An Information Systems Engineering Perspective

Edited by

Mohamed Ridda LAOUAR and Sean B. EOM

CAMBRIDGE SCHOLARS

PUBLISHING

Business Intelligence and Mobile Technology Research:
An Information Systems Engineering Perspective
Edited by Mohamed Ridda Laouar and Sean B. Eom

This book first published 2014

Cambridge Scholars Publishing

12 Back Chapman Street, Newcastle upon Tyne, NE6 2XX, UK

British Library Cataloguing in Publication Data
A catalogue record for this book is available from the British Library

ISBN (10): 1-4438-5507-3, ISBN (13): 978-1-4438-5507-5

TABLE OF CONTENTS

Part III: Miscellaneous Topics

PREFACE

In 2013, analytics and business intelligence, mobile technologies, and cloud computing were the top 3 priorities in the area of business and technology, according to a Gartner global survey of top 10 Chief Information Officers (CIOs). CIOs from 36 industries in 41 countries anticipate these information technologies to be a disrupting information technology that unexpectedly displaces an established technology over the next 10 years. Business intelligence systems consist of two major components, the data warehousing environment and the analytical environment. The latter is often further classified into two categories: reporting tools (dashboards and scorecards, reporting and query tools, etc.) and analytical tools. The analytical tools include decision support systems, executive support systems, online analytical processing (OLAP), and data mining.

All business organizations strive for increasing their growth by seizing new opportunities, reducing enterprise costs, attracting new customers and retaining old customers. In doing so, business intelligence and analytics allow business organizations to make better plans, informed decisions, and monitor their progress towards planned goals and objectives. The more disruptive power of IT technologies comes synergistically. Individual IT technologies do not work in isolation. Business intelligence systems are built on other digital technologies, such as mobile and collaborative technologies, cloud computing, virtualization, and enterprise resource planning and enterprise information systems. All of these IT technologies in combination bring forth their highest troublemaking forces to disrupt an existing market and seize new opportunities.

The Objective of this Book

This volume presents sixteen of the most insightful research papers amongst the various contributions accepted for presentations at the International Conference on Information Systems and Technologies (ICIST 2013) and International Conference on Software Engineering and New Technologies (ICSENT'12), held in Tangier, Morocco and Hammamet, Tunisia respectively. Two other papers originated from *AMCIS (*Americas Conference on Information Systems) *2012 Proceedings* *and Human Systems Management.* These papers truly represent what today's CIOs see as the top priority disruptive IT technologies that will help business organizations seize digital opportunities to increase their growth and reduce operating costs. The effective design, development, and management of enterprise information systems require multi-disciplinary efforts from management information systems (MIS), information system engineering (ISE), information science, and other disciplines. This volume showcases exemplary ISE research in business intelligence and mobile technology. It is our hope that the sharing of these works with the IS community in other disciplines will help develop more synergy across areas. ISE aims to design, develop, and manage enterprise information systems. Recent years have witnessed a steady growth of ISE research in the IS discipline, and it has become an important theme at major IS conferences. Cross-disciplinary research in ISE is also increasing.

The Structure of this Book

The book has three sections: Business Intelligence, Mobile Technologies, and Miscellaneous Topics. The first section of the book introduces readers to emerging topics in **business intelligence (BI)**. BI systems consist of three subsystems: data warehouses, business analytics, and performance management systems. BI systems enable users to access the data in data warehouses in real time, process the data to generate on-demand reports using online analytical processing, create decision alternatives using decision support systems/executive support systems, and predict future outcomes using data mining. BI systems utilize a wide range of management science and statistical models as well as visualization tools such as geographical information systems. Due to the breadth of BI topics, section 1 is further divided into three subtopics: decision support systems, data mining, and intelligent systems.

The decision support systems subsection includes the topics of negotiation support systems, decision support systems, and geographical information systems. Abroun and others propose a new concept and architecture of integrating wireless sensor networks within an enterprises information system into corporate data warehouses. They further discuss the benefits and different challenges for integrating wireless sensor networks into data warehouses. The next paper in this section presents a combination of psychological and rational approaches to simulate a complex negotiation behavior in the bilateral automated negotiation process. Autonomous agents could be intelligent software entities, distinguishable from a psychological character (such that conciliatory, neutral or aggressive), and a rational reasoning to take part in the process of searching gain with eventual agreement.

The third paper in this subsection deals with the empirical testing of DeLone and McLean's IS success model in the context of a Business Intelligence (BI) system in a police organization in Sweden. The last paper in decision support system, modeling knowledge management systems for component-based software engineering, addresses modeling knowledge management systems for the component-based software engineering CBSE domain by using a popular knowledge Engineering technique.

The second subsection presents two chapters that focus on data mining. Data mining is the process of discovering knowledge in large databases. It aims to find meaningful insight and relationships among attributes in the large database using a wide range of tools and techniques in statistics, management science, information systems, and artificial intelligence. To derive previously unknown relationships and insight, data mining uses tools and techniques for the following.

- Prediction
- Classification
- Segmentation/clustering
- Affinity analysis using association rules.

These data mining techniques can be broadly categorized into two groups: unsupervised and supervised. Ameur et al. present a method of segmenting satellite images based on the vectors of the gray level. Their work proposes a new approach to segment satellite images, based on the statistics of ranks to distribute the image in homogeneous regions. This algorithm takes into account the relation which binds the spatial positions and the ranks of the gray levels in a given neighborhood. This study highlights the efficiency of the characterization of regions by means of the statistics of ranks of the grey level. To deal with the same line of research to develop data mining algorithms that aim to minimize the number of frequent item set, Selmi, Gammoudi, and Harrathi propose a Prime Number method for improving the performance of formal concept extraction algorithms. The use of prime numbers allows one to perform the run time of two operations: intersection and union between sets considering the assertion that the comparison of numbers is faster than the comparison between strings.

The third subsection of Section 1 addresses intelligent systems. Intelligent systems utilize a wide range of artificial intelligence tools and techniques to develop information systems that can mimic a human-being.

Intelligent systems have been developed to make information systems/computers/devices that use human knowledge and reasoning to solve problems (expert systems). Furthermore, intelligent systems are developed to understand human language (natural language processing systems), and human spoken language (speech understanding/ recognition systems), perform a variety of manual tasks (robotics and sensory systems), and see and recognize objects (computer vision recognition systems).

Intelligent systems have been designed with several other advanced techniques such as the following.

- machine-learning techniques
- case-based reasoning
- genetic algorithms
- fuzzy logic and fuzzy inference systems
- support vector machines
- intelligent agents
- integrated advanced systems, which use a combination of individual intelligent tools.

Fuzzy neural networks: Neuro-fuzzy modeling (or fuzzy neural networks) is an example of integrated advanced systems. It refers to the method of applying the various learning techniques developed in the neural network literature to fuzzy modeling or to a fuzzy inference system (FIS). The basic structure of a FIS consists of three conceptual components as follows.

- A rule base, which contains a selection of fuzzy rules
- A database which defines the membership functions (MF) used in the fuzzy rules
- A reasoning mechanism, which performs the inference

Nahraoui, et al. developed a fuzzy neural network to predict the acoustic form function *(FF)* for an infinite length cylindrical shell excited perpendicularly to its axis. The Wigner-Ville distribution *(WVD)* is used like a comparison tool between the FF calculated by the analytical method and that predicted by the neuro-fuzzy techniques for a copper tube. This fuzzy logic system is able to predict the *FF* with a mean relative error *(MRE)* about 1.7%.

SVM and vision recognition system: Gattal and Chibani present another interesting example of an integrated advanced intelligent system that combines support vector machines and a computer vision recognition system. Gattal and Chibani are interested in a particular and very serious aspect in the domain of automatic recognition of normalized and not normalized isolated handwritten digits. The latter concerns feature generation, which requires the collection of a large database of digits. As the objective is to improve the performance of our recognition system, they are focused on combining different pertinent features from the digits.

The steps of the proposed approach are as follows. First, they start by applying different structural feature generation methods which are extracted from each image of handwritten digits. Furthermore, they use the uniform grid technique to increase the recognition rate. Then, the classification is done using SVM. The proposed approach was tested, and the results achieved are very encouraging.

Machine Learning: Machine learning has become one of the major preoccupations of computer science and artificial intelligence. Members of commonality learning face the significant challenge of giving machines the ability to learn and generalize. However, these algorithms are participating in its structure to organize data, select, and change the type of kernel function implemented. First, feature selection can retain only the relevant attributes, such as maintaining performance. On the other hand, the choice of the kernel function and parameter settings affects this performance.

Youcef and Yamina provide a rigid theoretical basis of machine learning algorithms. They began their study in this context, and they were faced with a conundrum in respect to the wide range of machine learning techniques proposed in the literature. How does one create a generic structure for modeling a learning algorithm inspired by the queuing system to implement a dedicated system, for example the classification of medical images? This system allows the aggregation of the features or instances of learning/test that are out of the original set.

Genetic Algorithms: The Web services technology offers great potential to overcome the problems of interoperability. It is a promising framework for the integration of applications and for managing interactions between various partners in a distributed environment, which is the Web. In the basic infrastructure around the SOAP standard, WSDL, UDDI is sufficient to establish interoperable and integrated component. However, it is insufficient to make automatic and efficient multiple tasks at lifecycle of web services, for example, the composition and also the discovery of the required services.

Bekkouche, et al. focus on web service composition, where a set of services is composed to satisfy the user request, based on the semantic descriptions of web services and their non-functional properties (QoS parameters). The user may also express its requirements through non-functional constraints (QoS constraints). The purpose of their work is to present an approach for web services composition based on a genetic algorithm that focuses on two aspects in order to find the optimal solution: the functional and non-functional aspects.

The second section of this book is titled, Mobile Technology. Mobile technology refers to the technology for wireless computing devices as well as wireless networking. The wireless devices include mobile phones, tablet computers, GPS navigation devices, etc. Mobile computing devices work with wireless networking technologies including virtualization and cloud computing. The first paper in this section discusses a key enabling technology, load balancing (LB), for virtualization and cloud computing. Cloud computing allows the user to virtualized network access to a service. It is load balancing that distributes and optimizes service requests to network resources using a scheduling algorithm. El Morabet et al. provide an overview of self-organization network (SON) mechanism. Load balancing in SON has been recognized as an effective means to increase network performance. The authors propose an algorithm which resolves the LB problem in the Long Term Evolution network.

The second paper, titled Cloud as a Major Catalyst of Change in Contemporary Business Evironment, sheds light on the use of Service-Oriented Architecture (SOA) to design the services deployable in the Cloud. This is done using business process reengineering to simplify applications by both internal and external users of the organizations, which are interested to move to Cloud computing. The use of SOA in the Cloud provides services data abstraction. This abstraction can be understood as services messaging metadata, which can be in the shape of XML, XSD or other set industry standards providing interoperability to a user's request. SOA eases the development of ever-changing applications that compare data with stationary applications while maintaining a decoupled relationship between these applications on a simultaneous basis.

The third paper, titled Security Attacks in Mobile Ad Hoc Networks, provides a comprehensive survey of the existent security threats that mobile ad hoc networks (MANETs) face. Abdelaziz, Nafaa, and Salim investigate selfish misbehaviors and malicious attacks through the different TCP/IP layers and discuss how these attacks affect the network. MANETs are wireless multihop networks that are dynamically constructed by autonomous mobile nodes without the support of any infrastructure or centralized administration. This new paradigm of wireless communication aims to make communication possible in some situations where the services offered by both wired networks and WLAN are unavailable. However, the characteristics and structure of MANETs make them prone to attack using different techniques often used against wired networks as well as new methods specific to MANETs.

The fourth paper in this section is titled, "A Smarter Way to Implement Circuit Switch Fallback and Its Modified Call Flow to Reduce the Call Setup Time." The long term evolution (LTE) is the newest mobile technology, dedicated mainly for high-speed data transfer. Because the LTE uses packet switching, voice calls in the global system for mobile communication, and universal mobile telecommunications system use circuit-switching technology, Operators/Carriers will have to take on adequate solutions to overcome the hurdle with the adoption of LTE. They include Voice-Over LTE, Circuit Switched Fallback (CSFB), and Simultaneous Voice and LTE. Jouihri and Guennoun propose a smart solution to implement CSFB as well as interesting recommendations to reduce the call setup time, which will result in a more comfortable user experience.

The last paper in this section is titled, An Architectural Framework to Design a Healthcare Information System. It presents an architectural framework for guiding the design of such systems. Following the best practices and guidelines of the most notable web developments processes, the framework in this chapter takes into account the recent advances in the Web and mobile technologies. Like any other information system in organizations, healthcare information systems are increasingly benefiting from recent advances in mobile technology and wireless networks. It seems to be self-evident to design a system that would support both patients and their healthcare providers in the process of treatment.

The third section of the book includes three chapters that discuss miscellaneous topics, such as ontology evaluation and e-learning. Ontology evaluation, from a logical and empirical perspective, has been recognized as a critical phase in an ontological engineering research field. Most well-known ontology development methodologies include evaluation as their part, either at the end or through the whole ontology development process. However, the profusion of methodologies has made it difficult for researchers to develop generic approaches for evaluating the ontological quality that are independent of any specific methodology. One of the most well-known ontology assessment approaches is the one achieved by humans, who attempt to assess how well the ontology fulfills its predefined criteria, standards and requirements. Chapter 16 by Zemmouchi-Ghomar and Ghomari propose a methodological baseline for ontology assessment by humans (domain experts and end-users) via a questionnaire. It aims to support a component-based evaluation.

The next chapter in the third section, Ontology–Driven MVC: A Variant of MVC architectural style by Chaturvedi and Prabhakar, highlights the drawbacks of the MVC architectural style and proposes an ontology driven solution to address these problems. Despite the separation of concern provided by the MVC architectural style, some inter-component dependencies, which constitute the liabilities of the style, still remain. In this paper, the proposed solution of the ontology is used as an active run time component. The ontology is generic and can be used for any application that is based on MVC style. The authors demonstrate the usage of this style through an example implementation based on both the traditional MVC style and the ontology–driven MVC style.

The final chapter by Eom et al. empirically tested the DeLone and McLean model of information systems success model in a university e-learning context using structural equation modeling. The results indicated that system and information quality both affected the two mediating constructs, system use and user satisfaction. These, in turn, influence the e-learning systems success. The nature of e-learning systems is fundamentally different from the environment in which the DeLone and McLean model is developed and tested, either voluntary or quasi-voluntary environments. However, e-learning systems are to be used in a mandatory environment. For that reason, the DeLone and McLean model has a limited explanatory power for explaining the role of e-learning systems on the outcomes of e-learning. Consequently, we conclude that the primary focus of the empirical research of the e-learning system should shift from the system factor toward human and design factors. E-learning systems empirical research is focusing on the critical success factors of e-learning outcomes that are comparable to or superior than traditional face-to-face learning system.

The Audience of this Book

This book is for practitioners, researchers, managers, and graduate students in computer science, information systems, information science, and computer engineering. Moreover, libraries in university, profit and non-profit organizations around the world will be potential customers. We believe that information systems researchers within the other disciplines will benefit from more dialogues and collaborations with each other. Greater synergies will promote and further advance the IS field as a whole.

PART I

BUSINESS INTELLIGENCE

CHAPTER ONE

TOWARD THE INTEGRATION OF WIRELESS SENSOR NETWORKS INTO AN ENTERPRISE INFORMATION SYSTEM'S DECISION-MAKING MODEL

O. ABROUN, A. TAHIRI, N. AKNIN AND K. E. EL KADIRI

Wireless sensor networks (WSN) are a new technology that opens new opportunities for data retrieval, and with its different and multiple real-world applications it takes information access to an advanced level. Due to the importance that information has today, benefits that WSN brings to information technologies might be described as overwhelming.

Moreover, within the world of enterprise information (EIS), information has a special value as the backbone of daily decision making systems. The importance of information in decision making represents a motivation that pushes organizations to the continuous development of their decision models.

Accordingly, the progress of both WSN and Information Technology (IT) within the context of EIS inspired us to suggest a new model that aims to integrate WSN within EIS. However, we believe that such integration is not free of challenges.

In our research, we introduce the main concept of WSN integration within the EIS decision model, and we discuss the different challenges that face this incorporation.

Introduction

Information and information technology witnessed a very significant development recently, especially in the Enterprise world, where they (Information and Information Technology) have a huge influence on

Enterprises structures and play an important role in their development. For this reason, we consider Enterprise a very sensible field where information becomes very important, especially in decision making.

On the other hand, a source of very valuable information that we find still is neglected in the enterprise world has made an explosive evolution during the last decade, i.e., a set of small, autonomous devices working together to solve different problems named Wireless Sensor Networks (WSN).

Based on this powerful evolution, we aim to discuss how WSN can be integrated to EIS, and also to study the difficulties that this kind of incorporation may face.

To achieve this goal, we first discuss the main concepts of information systems, and their role and influence on the EIS structure. Then we introduce WSN technology basics. Finally we discuss the different challenges to be faced when WSN technology is integrated to the world of EIS.

Information Systems and their Impact on Organizations

Davis (2000) defined the components of information systems:

> The Information System or Management Information System of an organization consists of the information technology infrastructure, application systems, and personnel that employ information technology to deliver information and communications services for transaction processing/operations and administration/ management of an organization. The system utilizes computer and communications hardware and software, manual procedures, and internal and external repositories of data. The systems apply a combination of automation, human actions, and user-machine interaction.

In addition, the importance of information systems can be seen in its influence on many levels, for example in the sensitive field of stock markets where it drives different transactions. We believe that the development of every system today hardly is related, on many levels, to the use of information technology.

Furthermore, as specified by Mousavi and BadrAbady (2008), the influence of information systems on EIS can be on the following levels:

- Operations: improves the efficiency of the main business operations, regardless their nature.
- Transactions processing: manages different transactions.

- Decision support: provides relevant information for decision making.
- Employees monitoring and performance evaluation: offers the possibility to follow employees' presence and performance.
- Documentation and Communication: conserves the main work status of business.

Therefore, we see that an organization's success is dependent on many levels on the use of information technology, and its influence is overwhelming, especially on the decision-making level.

Wireless Sensor Networks

WSN can be defined as a set of small, autonomous devices working together to solve different problems. It is considered to be a new technology that has been experiencing significant growth during the last decade. As time passes, people are becoming more aware of the important role that those microcontrollers may play in many aspects of their lives.

Navneet Kaur (2013) defines WSN major components as:

> **Sensor Field**: An area in which the nodes are placed.
>
> **Sensor Nodes**: Sensor-nodes are the heart of the network. They perform collecting and routing of data back to a sink.
>
> **Sink**: A sink is a sensor node with the specific task of receiving, processing and storing data from the other sensor nodes. It serves to reduce the total number of messages that need to be sent; hence reducing the overall energy requirement of the network. The network usually assigns such points dynamically. Regular nodes can also be considered as sinks if they delay outgoing messages until they have aggregated enough sensed information. Sinks are also known as data aggregation points.
>
> **Task Manager**: The task manager also known as base station is a centralized point of control within the network, which extracts information from the network and disseminates control information back into the network. It also serves as a gateway to other networks, a powerful data processing and storage centre and an access point for a human interface. The base station is either a laptop or a workstation. Data is streamed to these workstations either via the internet, wireless channels, satellite etc. So hundreds to several thousand nodes are deployed throughout a sensor field to create a wireless multi-hop network. Nodes can use wireless communication media such as infrared, radio, optical media or Bluetooth

for their communications. The transmission range of the nodes varies according to the communication protocol in use.

Thanks to this easy-to-deploy infrastructure especially since it does not have a strict topology (instead of being ad hoc), WSN can be deployed everywhere. Consequently, we consider the world of EIS to be full of precious data-retrieval opportunities. Also we believe that WSN integration into EIS is a very useful process that is full of advantages.

WSN Integration Within Enterprise Information System

EIS is composed of many connected information-based systems. The idea of integrating WSN to this information system is founded on adding a new member to the EIS body, or even feeding base systems with raw data coming from the outside world, in order to prevent accidents, to wisely manage human resources, to allow a data-based warehouse remote control, to monitor products quality and for many other practical applications.

Based on those potential applications on the one hand, and the different advantages that WSN have as a new data source on the other, we assume that EIS will take the following form after integrating WSN:

Figure 1: The Suggested Information System Model

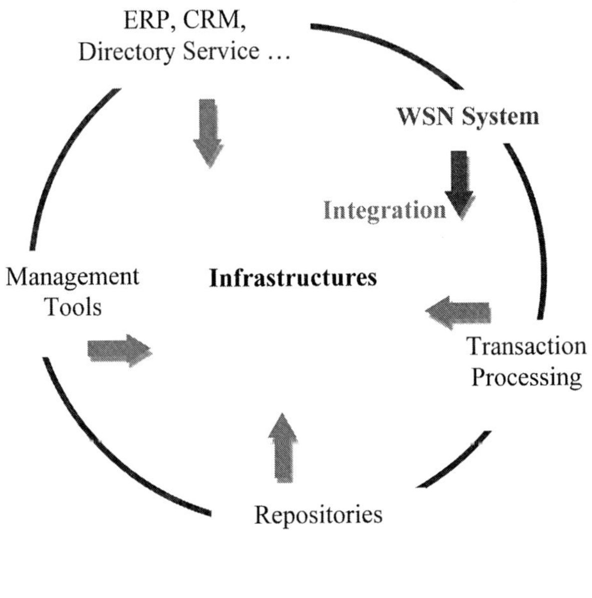

As indicated on the above figure, the different components of an information system, as previouslydefined, will have a new member represented by WSN systems. By communicating to those components, WSN are going to feed the system with data about the outside world (e.g., temperature, humidity, CO_2 levels, etc.). However, as the addition of a new member to an existing system may result in different changes. The integration of WSN into EIS presents many challenges. These challenges are discussed in the next section.

Integration Challenges

Based on the Middleware concept definition discussed by Gomez et al. (2008), we find that adding WSN to the EIS body needs to consider the following principles:

Heterogeneity

Every manufacturer in the market of wireless sensors provides data in its own format. As a result, the multitude of data formats is an important challenge to be faced, especially when EIS applications need to acquire data in a unified format.

Interoperability

The Enterprise Information System contains a large number of applications, which may be working in different environments. Consequently, WSN data should be accessed in the same way and with the same quality, regardless of the access environment.

Security

Information in general, especially when transported between two separate sides, needs to be secure in order to guarantee a high level of data trust. In the EIS world, information is very sensitive, so security is also another important challenge that the integration of WSN within EIS must overcome.

On the other hand, when data is available, another issue needs to be addressed when integrating WSN into the information system – the data access modes. A multitude of data modes is very important in the world of EIS, because the questions that need to be answered by decision-makers vary depending on the situation. Therefore we see that data collecting, which is the main function of WSN, should allow data access in the following modes:

- Raw Data Access:

The data, as collected by the sensors can be accessed without any filtering or preprocessing.

- Process Data Access:

Data, which after being processed by sensors, have the capability of executing such tasks, should be presented to the EIS.

- Historical Data Access:

Data history is one of the most important data access modes since many data mining techniques, e.g., prediction algorithms, are based on the collected data over time. Additionally, a degree of control over the system is needed in order to allow EIS administrations to configure WSN sensing based on their needs.

Therefore, the different challenges that WSN integration into EIS presents are reflected in the following figure:

Figure 2: Integration Model Middleware Architecture

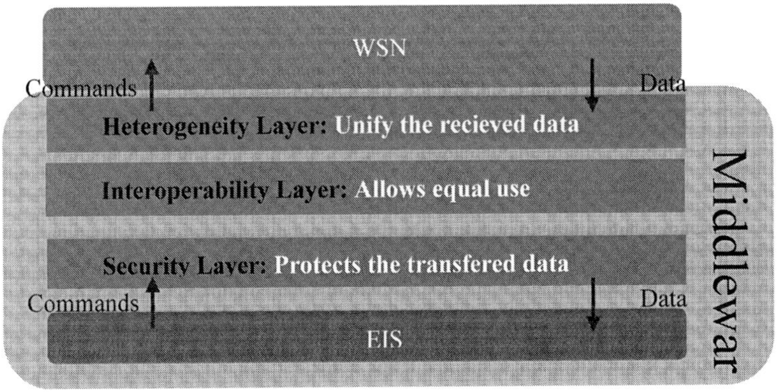

Conclusion

As we consider the process of incorporating WSN into EIS, we find that information systems in particular still is neglected by researchers. The different concepts discussed in this work will take decision making within EIS to an advanced level. However, due to the many challenges that such incorporation presents, bringing the benefits that WSN provide to the EIS world requires carefull planning. When done successfully, however, this process creates a framework that can open up a myriad of possibilities for increased productivity in any workplace that employs any variation of the EIS model.

References

Davis, G. B. (2000), "Information systems conceptual foundations: Looking backward and forward", *Organizational and Social Perspectives on Information Technology*, Vol. 41, pp. 61-82.

Gomez, L, Laube, A., and Sorniotti, A. (2008), "Design Guidelines for Integration of Wireless Sensor Networks with Enterprise Systems", *Proceedings of the 1st international conference on MOBILe Wireless MiddleWARE, Operating Systems, and Applications*, Article No. 12, pp. 12-15.

Mousavi, S.S. and Habiby BadrAbady, M., (2008) "The Role of Information Technology in Organizational Procedures' Improvement with Knowledge Based Approach-A Study of the Iranian Taxation Affairs Organization", *World Applied Sciences Journal* 3 Supple 2, pp. 55-56.

Navneet, sbeomKaur (2013), "Review on Load Balancing in Wireless Sensor Network", *International Journal of Advanced Research in Computer Science and Software Engineering,* Vol. 3, No. 5, pp. 1044-1047.

CHAPTER TWO

PSYRATIONAL STRATEGY TO SIMULATE NEGOTIATORS' BEHAVIOR IN BILATERAL AUTOMATED NEGOTIATION

DALEL KANZARI AND KHALED MELLOULI

The negotiation process aims to find an agreement between players (Fisher 2000) that satisfies their private requirements before reaching the negotiation deadline (Jennings et al. 2001). The latter have generally conflicting interests and try to acquire a compromise, even with incomplete information about the opponent's strategies, by using concession strategies. Many studies like the time-dependent strategy (Faratin et al. 1998), Bayesian learning (Zeng and Sycara 1998), reasoning based on constraints (Luo et al. 2003), and game theory (Binmore and Vulkan 1999), deal with agents' behaviors used to reach final agreements on matters of mutual interests. They focus only on one agent character that can influence the negotiation process and achieves results without combining diverse behaviors. Our motivation is to combine the following two characteristics: (a) the rational character that leads the agent to take into consideration the opponent's preferences when making a decision, and (b) the temperamental character that leads the agent to accept only the opponent's offer that ensures him a satisfied profit.

Our approach integrates psychological analysis (El-Nasr et al. 1998) (study of temperamental behavior during the negotiation process) and the rational decision-making process (study of reasoning to predict the adversary's strategy) in order to provide a methodological support to negotiation players.

Psyrational Negotiation Strategy

Given a round, composed of an offer and a counter-offer, and given the adversary's last action, the psyrational agent's reaction (choosing a suitable answer) is based on its internal "temperament" (conciliatory, aggressive, or neutral) and on its rational negotiation behavior. "Temperament" defines the psychological aspect of the agent's personality that plays an important role in determining the negotiation results:

- If the "temperament" is conciliatory, then the psyrational agent is more co-operative in reaching a final agreement.
- Otherwise, if the "temperament" is aggressive, the psyrational agent is more competitive and the negotiation might fail.

On the other hand, rational negotiation behavior aims to reach an agreement with mutual satisfied interests. The principle is to resolve the conflicts between competitors by considering the opponent's preferences in each decision. The model represents a win-win strategy modeled by a sequential game of decision-making.

Psyrational Agent Model

A psyrational agent is based on 4 interdependent subsystems organized in layers (see Figure 1):

- Communication and data representation: presentation of internal comprehensive data.
- Temperamental Behavior: study of cooperative, competitive, and neutral behavior.
- Rational Behavior: study of rational reasoning in decision-making process.
- Negotiation Behavior: study of the combination of Temperamental Behavior and Rational Behavior.

Figure 1: Psyrational Structure

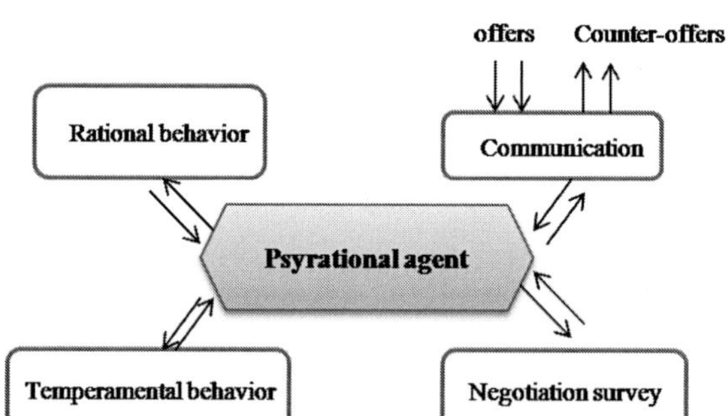

Rational Behavior

Rational Behavior leads the psyrational agent to take into consideration the other's aims and preferences when making a decision (Pruitt 1981, Osborne and Rubinstein 1994). It forms the collaborative side of the negotiation process, between players, in order to reach a final agreement with shared utilities.

Rational Behavior analyzes the complex problem of negotiation in several sequential sub-problems or sub-games composed of an offer and a counter-offer (Kanzari and Mellouli 2011) (See Figure 2.). A sub-game is a well-defined separate game that contains an initial node, and all the moves and information sets from that node remain in the sub-game. Each sub-game contains an equilibrium offer that can satisfy the both agents' goals.

Figure 2: Sequential Negotiation Strategies

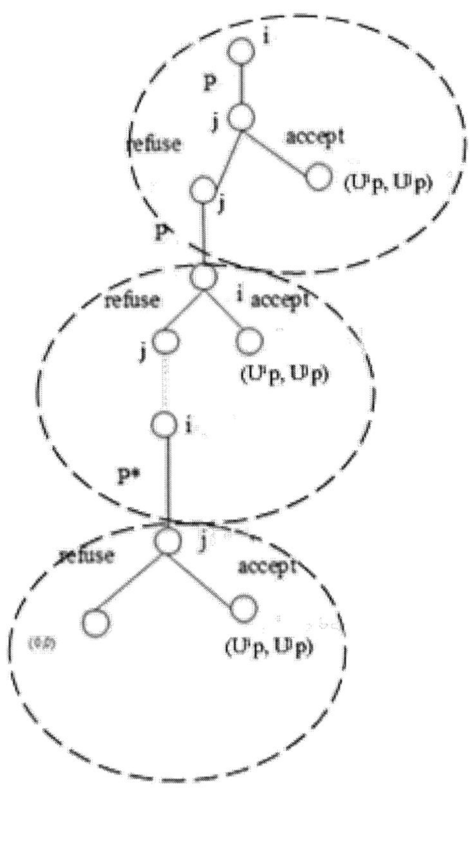

Figure 2 highlights the following notations:

- i, j: the players
- $U^i p$: the (i) player's utility by accepting the offer p
- $U^j p$: the (j) player's utility by accepting the offer p
- Each node represents a player that has two choices:
- Accept the adversary's offer (p) and then the negotiation process ends with the pair of utilities (U^i p, U^j p)
- Refuse the adversary's offer and propose a new offer if the negotiation deadline cannot be reached.

The aim of the rational approach is to resolve the conflict between players by finding the equilibrium offer that fulfils their preferences on each sub-game, before the negotiation deadline.

Temperamental Behavior

Temperamental Behavior reasoning integrates a fuzzy logic approach to control character variation influenced by the adversary proposal and to generate a related counter-proposal (See Figure 3.).

Figure 3: Temperamental Behavior

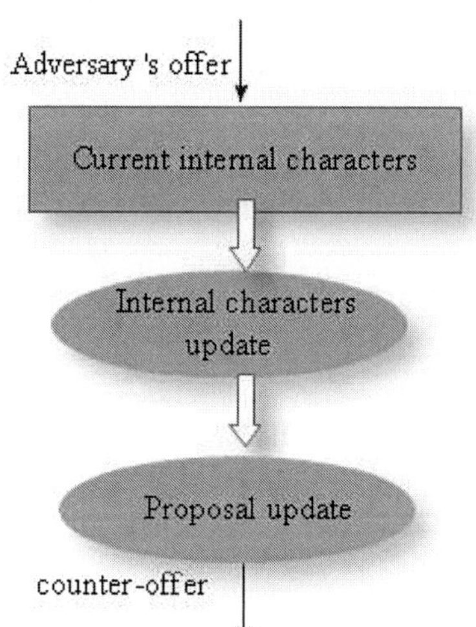

Given the last offer, and current psyrational agent's characteristic (conciliatory, neutral, and aggressive), the temperamental behavior model allows the update of these characteristics and the generation of an adequate counter-offer. This model emphasizes the influence of the negotiator's personality in sense of "temperament" in order to update the answers during the negotiation process. The characters' evaluation and the proposal updates are modeled by means of Mamdani's fuzzy systems (Liu 2002).

Negotiation Survey

The negotiation survey studies potential offers (Rosenschein and Zlotkin 1994) supplied from the temperamental reasoning and the rational reasoning, to elaborate the final system's decision making (See Figure 4.).

This module negotiation combines the outcomes of the strategies based on the agent's temperamental and rational reasoning to generate a personal-rational behavior endowed with both personal and rational characters.

Let P, defined in $P = \delta P_r + (1-\delta)^{\eta} P_t$ **(1)**, denotes the final counter-proposal:

$$P = \delta P_r + (1-\delta)^{\eta} P_t \quad \textbf{(1)}$$

P_r : the rational reasoning's output

P_t : the temperamental reasoning's output

$\delta \in [\,0\ 1]$: denotes the rationality degree:

- if $\delta \in [0.5\ 1]$, then the system's response is more rational.
- Otherwise, the system's response is more temperamental.

$\eta \in [\,0\ 1]$: denotes the degree of influence by the agent's personality:

- if $\eta \in [0\ 0.5]$, the agent's temperamental characters have great influence on the final decision making.
- Otherwise, the temperamental agent will have less influence on the final decision.

This module analyzes the two sub-systems (psychological and rational) results to find a suitable solution that satisfies the agent's objectives and guarantees a final compromise with the opponent before the negotiation deadline.

Figure 4: Negotiation Survey

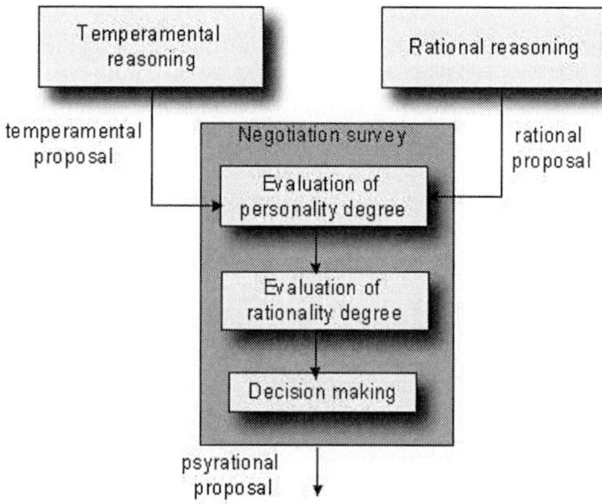

Experimental Evaluation

The aim of experimentation is to evaluate the psyrational strategy by comparing the latter to the time-dependent strategy and to create a simulation of a bilateral negotiation process between a psyrational buyer and a time- dependent seller (Faratin et al. 1998), for the price of a commodity.

The offer of a time-dependent seller is defined by (2):

$$P_s^t = \min_s + (1 - \alpha_s(t)).(\max_s - \min_s) \quad \textbf{(2)}$$

$\alpha_s(t) = (\frac{t}{T})^{\frac{1}{\beta}}$: The time-dependent function that determines the offer's concession speed according to the value of β, so if β <1, then the agent is aggressive. Otherwise, the agent will be conciliatory (Gratch and Marsella 2004, Raiffa 1982).

T: The negotiation deadline

mins: The minimal seller's price (reserved value)

maxs: The maximal seller's price

To evaluate agent strategies, we use the following measures:

The average utilities (Ui): denotes the intrinsic agents' benefits from negotiation playoff.

- Buyers' Utilities: $U_b = \max b - P^*$
- Sellers' Utilities: $U_s = P^* - \min s$

– p^* : The agreed price (accepted by the two parties).

Is some cases, if no deal is made then $U_b = U_s = 0$

The average number of rounds (Round_moy) to reach an agreement, is defined by $\text{Round\_moy} = \frac{\sum \text{rounds'numbers}}{\sum \text{sussful rounds}}$ **(3)**:

$$\text{Round\_moy} = \frac{\sum \text{rounds'numbers}}{\sum \text{sussful rounds}} \quad \textbf{(3)}$$

The agreement rate (Agr_rate), is defined by $Agr\_rate = \frac{\sum \text{sussful cases}}{\sum \text{all cases}}$ **(4)**:

$$Agr\_rate = \frac{\sum \text{sussful cases}}{\sum \text{all cases}} \quad \textbf{(4)}$$

These strategies (the psyrational and time-dependent strategies) are subdivided respectively into three tactics according to the agent's competitive, neutral or cooperative behavior (Table 1):

Table 1: Negotiation Tactics

Tactics	Name	Abbreviation	Parameters
Dep.time	Conciliatory	***C***	$\beta \in [0.01 \quad 0.2]$
Dep.time	Neutral	***N***	$\beta = 1.0$
Dep.time	Aggressive	***A***	$\beta \in [20.0 \quad 40.0]$
Psyrational	Conciliatory	***C***	$\delta \in [0.51 \quad 1]$ $\eta \in [0.51 \quad 1]$
Psyrational	Neutral	***N***	$\delta = \eta = 0.5$
Psyrational	Aggressive	***A***	$\delta \in [0.01 \quad 0.5]$ $\eta \in [0.01 \quad 0.5]$

To simplify the calculation, we limit the range of variable instances as follows: Tbmax(= Tsmax) ∈ [20 100] and mins = minb = 10.

Experiences

Experimentation focuses on the δ and η variation, and the players' behaviors (such as conciliatory, neutral, or aggressive). Table 2 shows the nine studied experiences.

Table 2: Negotiation Experiences

Experiences	Player's behaviors (buyers'/(against) sellers')
1	Aggressive psyrational buyers (Ab) / Aggressive times Dep. seller (As)
2	Aggressive psyrational buyers (Ab) / Neutral times Dep. seller (Ns)
3	Neutral psyrational buyers (Nb) / Aggressive times Dep. seller (As)
4	Neutral psyrational buyers (Nb) / Neutral times Dep. seller (Ns)
5	Aggressive psyrational buyers (Ab) / Conciliatory times Dep. seller (Cs)
6	Conciliatory psyrational buyers (Cb) / Aggressive times Dep. seller (As)
7	Conciliatory psyrational buyers (Cb) / Conciliatory times Dep. seller (Cs)
8	Conciliatory psyrational buyers (Cb) /Neutral times Dep. seller (Ns)
9	Neutral psyrational buyers (Nb) / Conciliatory times Dep. seller (Cs)

To distinguish between the influence of the temperamental and the rational approach on the negotiation process and the negotiation results, we focus on two examples of the negotiation simulation process between a psyrational agent and a time-dependent agent. The first example handles the temperamental behavior of psyrational buyers whereas the second treats the rational one (the value of δ and η determining the dominant approach)

Negotiation Simulation Process: Temperamental Agent (δ = 0. 25 and η = 1) against Time-Dependent Agent

Psyrational agents are less rational than temperamental ones (δ=0. 25). Figure 5 presents the average utilities of the temperamental psyrational agent (buyers b) and the time-dependent agents (sellers s). Additionally it represents the agreement rate of the negotiation process and the average rounds to reach agreement.

According to these graphs we notice that (a) the psyrational buyers' utilities are almost higher than sellers' utilities, (b) the agreement rate is nearly average, and (c) the average rounds to reach agreement are low.

We also notice that the agent's character influences the negotiation results. So if the psyrational agent is more aggressive, its average utility will degrade, the agreement percent will reduce and the average round to reach agreement will enhance.

Negotiation Simulation Process: Rational agent (δ=0. 6 and η = 0) against Time dependent Agent

Psyrational agents are more rational than temperamental. Figure 6 presents the average utilities of successful negotiation cases of rational buyers (Rb) against time-dependent sellers (s) as well as the agreement rates and, finally, the average rounds taken to reach an agreement.

Figure 6: (a) Average Utilities of All Cases (Successful and Failed), (b) Average Utilities of Successful Cases, (c) Agreement Rates, and Average Rounds

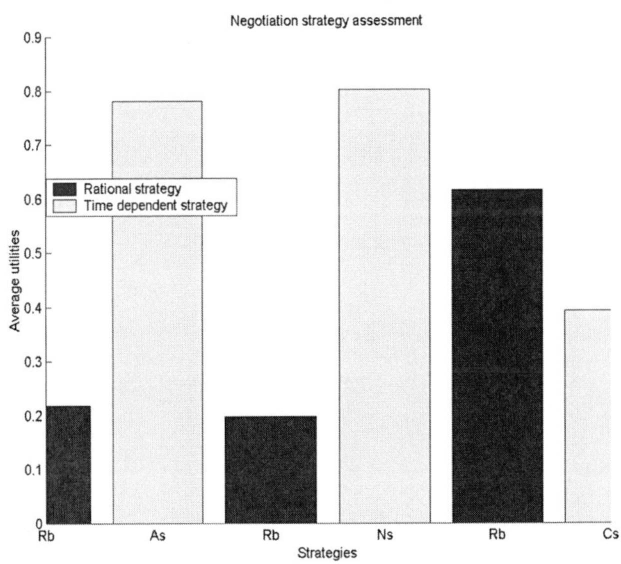

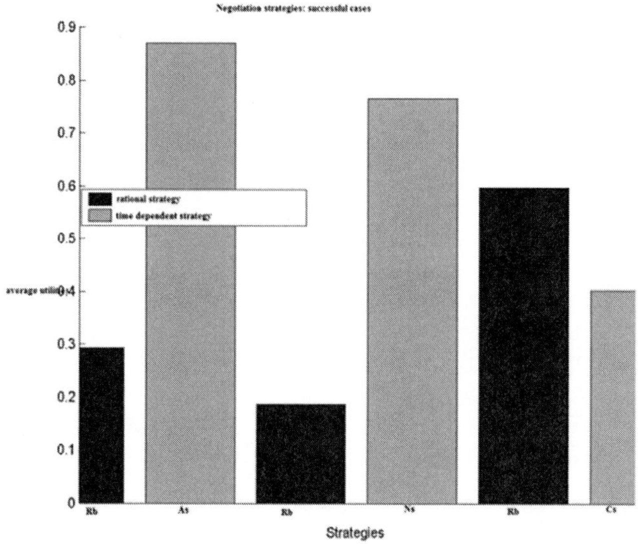

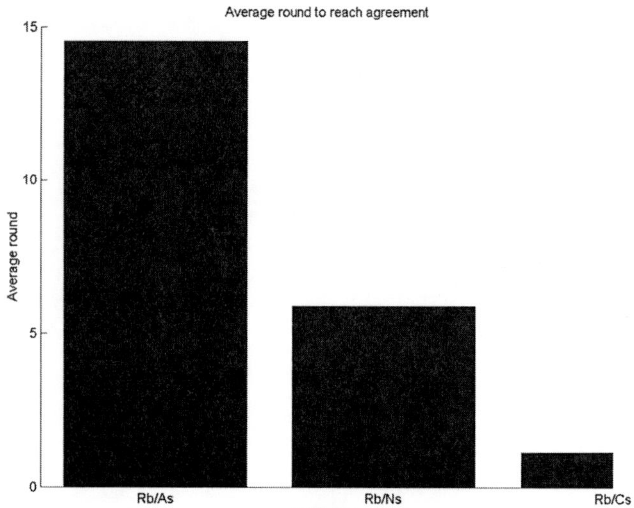

The experimental study of the psyrational strategy based on Rational Behavior (Rb) against the time-dependent strategy (sellers) shows a total success of all negotiation cases but with a little final gain for rational reasoning.

We noticed that the rational agent succeeds in resolving the conflict with its opponent but fails to reach gain.

Interpretation

We notice from the different cases that, if we increase the temperamental degree (η), psyrational buyer utilities become more performant than time-dependent seller utilities, but the average rounds to reach agreement will increase (time lost) and the agreement percentage will decrease.

Conclusion

This chapter studies a complex architecture, integrating intelligent techniques of decision-making systems to simulate agent negotiation behavior. The systems are based on fuzzy logic to model negotiator psychological aspects, and game theory to model rational reasoning throughout the negotiation process.

The combination of the psychological (temperamental) and the rational approach allows the psyrational agent to learn competitive, cooperative, or neutral behaviors, to adjust environment changes, and to ensure success in different circumstances.

References

Binmore, K. and Vulkan, N. (1999) "Applying game theory to automated negotiation." *Netnomics*, Vol 1, No. 1, pp. 1–9.

El-Nasr, M. S., Ioerger, T. R. and Yen, J.(1998) "Learning and Emotional Intelligence in Agents." *Proceedings of AAAI (American Association For Artificial Intelligence) Fall Symposium on Emotional Intelligence.*

Faratin, P., Sierra, C. and Jennings, N. R. (1998) "Negotiation decision functions for autonomous agents." *Robotics and Autonomous Systems*, Vol 24, No. 3–4, pp. 159–182.

Fisher, M. (2000) "Characterising Simple Negotiation as Distributed Agent-Based Theorem Proving – A Preliminary Report." *Proceedings of ICMAS00*, pp. 127–134.

Gratch, J. and Marsella, S. (2004) "A domain-independent Framework for modeling emotion." *Journal of Cognitive Systems Research*, Vol 5, No. 4, pp. 269-306.

Jennings, N. R., Faratin, P., Lomuscio, A. R., Parsons, S., Wooldridge, M. J. and Sierra, C. (2001) "Automated Negotiation: Prospects, Methods and Challenges." *Group Decision and Negotiation*, Vol 10, No. 2, pp. 199–215.

Kanzari, D. and Mellouli, K. (2011) "Rational strategy to perform Automated bilateral negotiation." *International Conference on Information Technology and e-Services – ICITeS'2011.*

Liu, P. (2002) "Mamdani fuzzy system: universal approximator to a class of random processes." *Fuzzy Systems, IEEE Transactions*, Vol 10, No. 6, pp. 756–766.

Luo, X., Jenning, N. R., Shadbolt, N., Leung, H.-f. and Lee, J. H.-M. (2003) "A fuzzy constraint based model for bilateral, multi-issue negotiations in semi-competitive environments." *Artificial Intelligence,* Vol 148, No. 1–2, pp.53–102.

Osborne, M. J., and Rubinstein, A. (1994) "A Course in Game Theory." *The MIT Press.*

Pomerol, J.-Ch. (1997) "Artificial Intelligence and Human Decision Making." *European Journal of Operational Research*, Vol 99, pp. 3–25.

Pruitt, D. G. (1981) "Negotiation Behavior". *Academic Press*, New York.

Raiffa, H. (1982) "The Art and Science of Negotiation." Cambridge: MA, *Harvard University Press.*

Rosenschein, J., and Zlotkin, G. (1994) "Rules of Encounter". Cambridge: MA, MIT Press.

Zeng, D. and Sycara, K. (1998) "Bayesian Learning in Negotiation." *International Journal of Human-Computer Studies*, Vol 48, pp.125–141.

CHAPTER THREE

AN EMPIRICAL TEST OF DELONE AND MCLEAN'S INFORMATION SYSTEM SUCCESS MODEL IN A PUBLIC ORGANIZATION[1]

OLGERTA TONA, SVEN A. CARLSSON AND SEAN B. EOM

Introduction

Business Intelligence (BI) systems provide, based on data analyses, relevant information to decision makers (Watson and Wixom, 2007). Many organizations implement BI due to its potentially significant impact on business performance (Hawking and Sellitto 2010, Watson and Wixom 2007). Although IT investments decreased in 2009, BI was still dominant on IT leaders' agendas by continuing to grow during 2010 as well (Garner 2009). In the 2011 SIM survey BI was ranked as the most important of all application and technology developments, and BI has been ranked as number one for the last three years (Luftman and Ben-Zvi 2011). In 2010, the revenue for BI, analytics and performance management increased by 13.4% compared to 2009, yielding around $10.5 billion revenue (Gartner 2011). Therefore, the success of BI systems is of significance to the organizations having invested in them.

It is important to point out that the implementation, management and evaluation of Information Systems (IS) differ between public and private sector organizations (Rosacker and Olson 2008). Public organizations are investing a lot of money in the implementation of different IS, but most IS evaluation research focuses on private sector organizations (Rosacker and Olson 2008). The relatively few IS evaluation studies done have, in

[1] First published in AMCIS 2012 Proceedings, (July 29, 2012), paper 10.

general, had a focus on information systems in health care or e-government.

There are many definitions of IS success and over the years different measures have been used to evaluate IS success. DeLone and McLean (1992) state that the success of a system is considered as a multidimensional construct difficult to define and measure. For this reason they presented their model, consisting of 6 measures, which could be used in evaluating the success of different information systems. This model, being the dominant IS success model, has been cited 4700 times (Google Scholar, April 24, 2012).

According to Iivari (2005), there is a lack of empirical tests of DeLone and McLean's model. Additionally, he suggests that empirical tests should be done with different types of IS and in different contexts. This paper will test empirically DeLone and McLean's IS success model in a public organization context and at the same time make a contribution to the sub-fields "BI success" and especially "BI success in police organizations". Generally, IS evaluation is concerned with the evaluation of different aspects of real-life interventions where IS are critical means in achieving the interventions' anticipated goals.

The remainder of the paper is organized as follows. The next section presents the IS success model used in the study. It is followed by a description of the research approach. Section 4 presents the results of the study, which is followed by the discussion section. Conclusions are presented in the final section.

The DeLone and Mclean Model

In order to organize the diverse research done on IS success factors over the years, DeLone and McLean (1992) proposed a model which presented a more integrated and comprehensive view of the success factors (Fig. 1).

In their model they showed that IS success does not depend only on one factor but on many of them, which are interrelated. This model consists of six dimensions: system quality, information quality, use, user satisfaction, individual impact, and organizational impact (DeLone and McLean 1992).

When the system is created it is characterized by system quality and information quality. The system is used by the users who can be satisfied or not with it, and during its use the system can have an individual impact. This impact in total will affect the organizational impact. System quality is regarded and classified in terms of technical level, and information quality in terms of semantic level. On the other hand, use, user satisfaction, and individual and organizational impacts are considered in order to evaluate the effectiveness of the system (DeLone and McLean 1992, DeLone and McLean 2003).

After 10 years, DeLone and McLean updated their work and presented a new model in 2003. They included as another dimension *service quality* and also combined the individual impact and the organizational impact in a new factor named *net benefits*. Additionally, the *intention to use* was added to the use dimension to complement it (DeLone and McLean 2003).

In this study, we have chosen to conduct our evaluation research using the DeLone and McLean model of the year 1992. Our intention is to study the effects of the system usage not only in relation to the whole organization but also the impact it has on its users in terms of their daily work. Therefore, the individual impact of BI should be measured separately from the organizational impact.

Figure 1: DeLone and McLean's IS success model

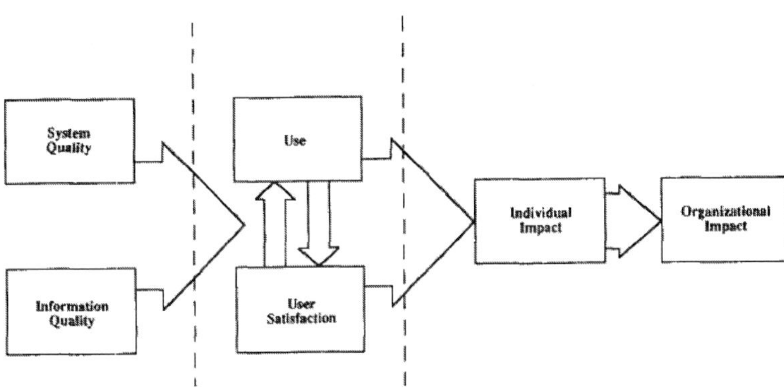

System quality is concerned with issues such as if there are 'bugs' in the system, the user interface, ease of use and sometimes even quality and maintenance of program codes (Seddon 1997). *Information quality* refers to the quality of the information which the system produces. It is considered an important factor when the system being evaluated involves the production of information to be used in decision making processes (Seddon 1997). Three different meanings of use have been presented by Seddon (1997). The first meaning relates *use* to the system benefits that shall flow from it; the second meaning defines *use* as a variable to describe behaviour; in the third meaning *use* is considered to lead to user satisfaction and individual and organizational impact as outcomes of a process model. In any case, from these three meanings, Seddon (1997) concludes that the first meaning is the appropriate one to be applied to Delone and McLean's model, where *use* shall be measured in terms of time and frequency of use. *User satisfaction* has been defined by Bailey and Pearson (1983, p. 531) as follows: "satisfaction in a given situation is the sum of one's feelings and attitudes toward a variety of factors affecting that situation". *Individual impact* as: "an indication that an information system has given a user a better understanding of the decision context, has improved his or her decision making productivity, has produced a change in user activity, or has changed the decision maker's perception of the importance or usefulness of the information system" (DeLone and McLean 1992, p. 69). *Organizational impact* measures the impact that the use of the system has on the overall performance of the organization (DeLone and McLean 1992).

Based on the DeLone and McLean model, the research model and the hypotheses related to this study are shown in Figure 2.

Figure 2: Research model and hypothesis to be tested

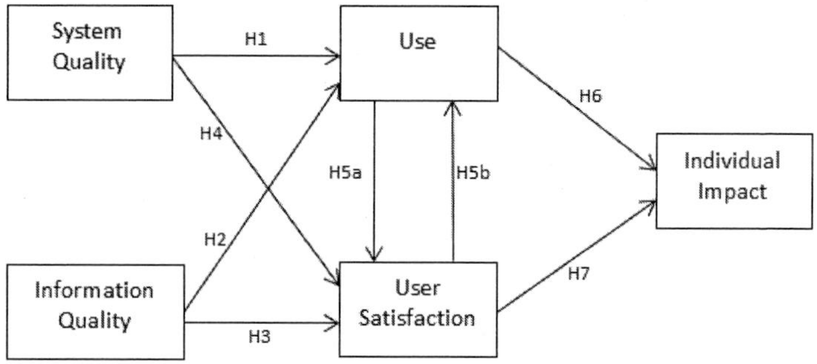

Figure 2, our research model, misses organizational impact. David and Jackson (2005) describe the difficulties in evaluating the benefits and costs related to systems in the crime area as they cannot be expressed in monetary terms. Furthermore, they suggest that a qualitative study would be appropriate to measure the potentially positive impacts and some of the potential costs. To comply with these suggestions, interviews with the key persons were found to be suitable to obtain the relevant information about the overall impact on the organization of using the BI system. For an elaborate discussion of the organizational impacts, see Carlsson et al.

(2010). Hence, the lack of quantitative data explains the exclusion of organizational impact from our research model.

Research Methodology

The Context of Study

A BI system operating in a public organization context will be evaluated using the DeLone and McLean model. The users involved in this study are employees in a police organization in Skåne, a southern region of Sweden. Skåne is the third largest police authority in Sweden and it has approximately 3240 employees, of whom approximately 2340 are police officers and 900 civil servants. The BI, created with the software QlikView, started as a single application based on the system RAR (a system for crime statistics, where all reported crimes are registered). QlikView is a BI/DSS software company, see: http://www.qlikview.com. The system was used by crime analysts to forecast when and where crime would occur. The system creates associations based on the processed data which make it easy to identify relationships between them. The information can be visualized by diagrams, tables or dashboards. The BI-applications within the organization exist mostly in two categories: "view" and "analyse". The view applications were available to all employees through the intranet, while the analysts have the same access as the viewers, but in addition they had the option to do more detailed analyses.

To collect data, both quantitative and qualitative approaches have been embraced. A questionnaire was constructed in order to measure five success factors: information quality, system quality, use, user satisfaction, and individual impact. The questionnaire was created in a program that the police possessed. Via a notifying e-mail, which contained a link to the questionnaire, the potential users in the police organization in Skåne were asked to participate in the study. Of approximately 3240 potential users (having diect access to the system), 367 questionnaires were obtained. Of the 367 respondents, only 28.1% (103) had used the system. In this study we were performing a BI system evaluation through the users' perceptions; therefore we were interested only in the 103 real users of the system. After the data cleaning and handling of the missing values, out of 103, 87 cases were used in our data model. In the absence of formative indicators in the model, the sample size required for this analysis is equal to ten times the largest number of structural paths directed at the dependent (endogenous) construct, "use" or "user satisfaction", in Figure 2 (Chin and Newsted 1999, Hair, Ringle, and Sarstedt 2011). Thus, the minimum sample requirement is 30. To establish adequate power of .80 (α = .05) in order for the effect to become significant, the required sample size is 76 (a medium effect size). With a larger effect size, it drops significantly to 35 (Cohen 1988, Green 1991). Therefore, our sample size of 87 is adequate to produce statistically significant results. Further interviews have been performed to address the issues and questions not covered by the survey.

The Questionnaire Design

Given the DeLone and McLean (1992) model, the aim of the questions in the survey was to measure its first five success factors. Each factor was measured by one or several items. Most of the items in Table 1 were based on other questionnaires that have been used, tested and validated in other research studies. Most of the items were measured using a 5-point Likert scale.

Data Analyses

The model is tested with Partial Least Squares (PLS), a structural equation modelling technique. Chin and Newsted (1999) argue that PLS is a widely used method in small and medium sample sizes. It models the structural and measurement paths. The former is related to the theoretical relationships among the latent variables and the latter is related to the relationship between a latent variable and its indicator. Therefore, it can be used for theory confirmation as well as testing the existing relationships. The model of figure 2 is tested using WarpPls 2.0 software. The existence of the mutual influence between use and user satisfaction leads to the testing of two models. Model 1 includes use as a predictor of user satisfaction, and model 2 includes user satisfaction as a predictor of use.

Measurement Model Estimation

In order to test the relationships in the structural model, first the validity and reliability of the measurement model shall be evaluated (Fornell and Larcker 1981). The composite reliability and Cronbach's coefficient alpha measured were above 0. 7 for most of the variables (Table 2), a threshold value recommended by Nunally and Burnstein (1994). Only 'use' shows a poor Cronbach Alpha as it is below 0.7. Additionally, average variance extracted (AVE) values were above 0.5 in all the variables, which means that the variance of the construct is larger than the error (Fornell and Larcker 1981). Overall, this study indicates a relatively good reliability of the measurement instrument used.

Construct validity is assessed through establishing both convergent and discriminant validities. Convergent validity refers to the extent to which a set of indicator variables loads together and it is demonstrated when the variables load highly (loading >0.50) on their associated factors. Individual reflective measures are considered to be reliable if they correlate more than 0.7 with the construct they intend to measure. Table 3 shows that most of the loadings were above 0.8 for the five constructs, higher than the threshold value 0.7. When indicator variables do not cross-load on two or more constructs, each construct is said to be demonstrating discriminant validity.

Table 1: Constructs and measures

CONSTRUCT	ITEMS TO BE MEASURED	Based on
SYSTEM QUALITY	Is BI easy to use? Is BI easy to learn? Are you satisfied with the response times of the system? (E.g. the time it takes to get an answer to a question)	Doll and Torkzadeh (1988) DeLone and McLean (1992)
INFORMATION QUALITY	**CONTENT** Does the system have the type of information you need? Does the system provide sufficient information? **ACCURACY** Are you satisfied with the overall accuracy of the system? **FORMAT** Is the output you get presented in a useful format? (eg. charts, tables, graphs)? Is the information presented in clear and understandable way? **TIMELINESS** Does the system provide up-to-date information?	Doll and Torkzadeh (1988)
USE	How often do you use BI? How much time do you spend when using the system?	Seddon (1997)
USER SATISFACTION	How do you rate your satisfaction with BI?	Rai et al. (2002)
INDIVIDUAL IMPACT	By using the system, you get a better understanding of important trends. By using the system you improve the quality of your work. You rate BI an important tool in your work.	DeLone and Mclean (1992)

Table 2: Latent variable coefficients

	System Quality	Information Quality	Use	User Satisfaction	Individual Impact
Cronbach Alpha	0.840	0.917	0.512	1.000	0.840
Composite Reliability	0.905	0.935	0.804	1.000	0.904
AVE	0.761	0.706	0.672	1.000	0.758

Table 3: Loadings and Cross-Loadings

	System Quality	Information Quality	Use	User Satisfaction	Individual Impact
sq1	**0.931**	-0.101	0.076	0.147	-0.077
sq2	**0.907**	0.027	0.026	-0.270	0.098
sq3	**0.770**	0.090	-0.122	0.140	-0.023
iq1	-0.298	**0.807**	0.097	-0.629	0.029
iq2	-0.225	**0.833**	0.093	0.017	0.123
iq3	0.126	**0.843**	-0.063	-0.099	0.166
iq4	0.100	**0.834**	0.011	0.412	-0.199
iq5	0.284	**0.859**	-0.024	0.013	-0.162
iq6	-0.008	**0.865**	-0.106	0.256	0.045
u1	-0.037	-0.210	**0.820**	0.357	0.202
u2	0.037	0.210	**0.820**	-0.357	-0.202
us1	-0.000	-0.000	0.000	**1.000**	0.000
ii1	-0.109	0.109	-0.085	0.305	**0.864**
ii2	-0.212	-0.017	0.017	0.129	**0.908**
ii3	0.342	-0.094	0.069	-0.455	**0.838**

Table 4: Construct correlation and square root of AVE on the diagonal

	System Quality	Information Quality	Use	User Satisfaction	Individual Impact
System Quality	**0.872**				
Information Quality	0.715	**0.840**			
Use	0.338	0.199	**0.820**		
User Satisfaction	0.737	0.870	0.222	**1.000**	
Individual Impact	0.626	0.697	0.425	0.692	**0.871**

In PLS, discriminant validity was assessed using two methods. First, by examining the cross-loadings of the constructs and the measures (Table 3); Second, by comparing the square root of the average variance extracted (AVE) for each construct with the correlation between the construct and other constructs in the model (Fornell and Larcker 1981, Chin 1998).

Results

In this study, 2 models are analyzed based on DeLone and McLean (1992). In the first model, use affects user satisfaction and in the second model user satisfaction affects use. The PLS results are shown in Figure 3. The values of β_1 correspond to Model 1 and β_2 to Model 2. The same logic applies to R^2 where R^2_1 refers to Model 1 and R^2_2 refers to Model 2.

Figure 3: Results of the analyses

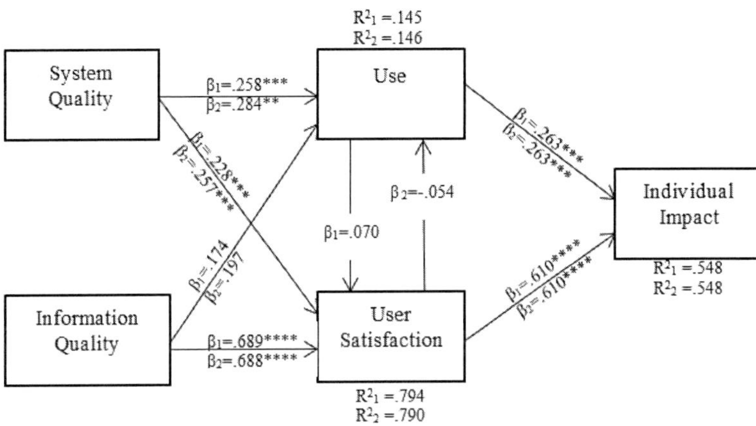

**** $p < .001$, *** $p < .010$, ** $p < .050$, * $p < .100$.

The results show that the variance of Individual Impact is explained 54.8% in both models while the variance of User Satisfaction is explained 79.4% in Model 1 and 79% in Model 2. Figure 3 indicates that System Quality is a good predictor of both Use and User Satisfaction with $p < 0.01$. Information Quality is a strong predictor of User Satisfaction ($p < 0.001$) in both models, but not of Use. Use ($p < 0.01$) and User Satisfaction ($p < 0.001$) are good predictors of Individual Impact in both models. Additionally, Use is not a significant predictor of User Satisfaction and vice versa.

Discussion

Based on the structural analyses, the findings of this study regarding the hypotheses are shown in Table 5.

Table 5: Hypotheses

	Hypothesis	Model 1	Model 2
H1	System Quality is positively associated with Use	Supported	Supported
H2	Information Quality is positively associated with Use	Not Supported	Not Supported
H3	Information Quality is positively associated with User Satisfaction	Supported	Supported
H4	System Quality is positively associated with User Satisfaction	Supported	Supported
H5a	Use is positively associated with User Satisfaction	Not Supported	–
H5b	User Satisfaction is positively associated with Use	–	Not Supported
H6	Use is positively associated with Individual Impact	Supported	Supported
H7	User Satisfaction is positively associated with Individual Impact	Supported	Supported

A higher System Quality leads to a higher Use of the system (H1), a result similar to the study by Iivari (2005). Relating this fact to the context in which this system is evaluated we conclude that the employees of the public organization will use the system more if it is easy to use and learn. It is worth mentioning here that most of the employees found the BI system easy to use. As our interview respondents described, it is only a matter of pushing buttons and selecting, and then one immediately gets access to the data and information that is being looked for.

We have not found a significant relationship between Information Quality and Use (H2). This means that higher information quality does not necessarily lead to higher usage of the system. This result is supported by the studies of McGill et al. (2003) and Iivari (2005), where the path from Information Quality to Use was not significant. Based on the questionnaire and the interviews conducted, most of the users seemed satisfied with the format and timeliness of the system, while improvements were required for the content and the accuracy (Carlsson et al. 2010). Although overall they seemed satisfied with information quality, they provided further comments about the difficulties they faced when trying to interpret the data produced by the system. This might be a result of their lack of knowledge of the legacy systems on which the BI system is built. Additionally, a respondent said that "*when you come deep down in the analyses you must know every part of it and understand the complexity of the system of course and if you don't ... well, you can push some buttons and get an answer but if you don't know what you have pushed you don't know what kind of answer you'll get of course.*" We believe that the understanding of the data (a good knowledge of the legacy systems) in a BI system plays an important role in BI usage from the employees; therefore, we believe that it should be included as a variable in the other BI system evaluations.

User Satisfaction is predicted by both System Quality and Information Quality (H3, H4). These results support Seddon and Kiew's (1994) study, where Information Quality and System Quality were positively related to User Satisfaction. The relations between these constructs are commented on by the users. Some users have specified different reasons why they have not rated this item higher. Some are not satisfied with the updates; they find difficulties in understanding the data which are produced by the system. Others believe that their needs are not fully supported; they feel somewhat insecure about the reliability and correctness of the data and that they lack knowledge of the main purpose and functions of the system. On the other hand, they seemed relatively satisfied with the ease of use, ease of learning, the presentation of the information and the update of the system. Hence, we deduce that the higher system quality and information quality is, the higher user satisfaction will be.

The research showed that neither Use nor User Satisfaction is a predictor of the other; hence both hypotheses H5a and H5b are not supported. The study of Roldan and Leal (2003) showed insignificant paths from User Satisfaction to Use and concluded that other predictor variables for IS use should be identified. Use is predicted only by System Quality, but neither by Information Quality nor User Satisfaction. We believe that there can be other factors influencing BI use in a police organization. Davis and Jackson (2005) have discussed the concept of slow diffusion of information concerning different systems in police organizations and according to Pick (2008), the more obvious the benefits of a system become for the organization and users, the more the usage rate will increase. It seems that the advantages of using this system have not reached the potential users as they lack knowledge about it. Additionally, in the organization, there are identified two main user groups: the analysts and the viewers. Based on the interview respondents, the users who have view access level usually use the system to get an understanding of the organization's performance and also use it for production and monitoring, while the analysts perform more detailed analyses for crime solution and crime prevention. In addition to the interviews, the Chi-square test indicated that BI usage was significant depending upon the access level (Carlsson et al. 2010). Also, within the viewers we can distinguish another group, the managers, who have the possibility of better control over personnel and an enhanced way to measure the work of the individual.

Hence, in a BI system evaluation we believe that other factors such as: the positions inside the organization related to the work the employees perform, the BI access levels, knowledge about the legacy systems leading to a better understanding of the data, and the information that the users have about the advantages and benefits of the system may affect BI usage within the organization.

In contrast to the findings of Iivari's (2005) study, Use is positively related to Individual Impact (H6). DeLone and McLean (2003) during a literature review reported that this relationship was positive in most cases where system use was typically voluntary. Iivari (2005) explained that the rejection of this hypothesis in his study might have been partly due to the mandatory nature of the system. In our case we can categorize it as a voluntary system, where the users consider the system as a resource of information about general issues, which they choose to access during their working process. By using the system, the users have a better understanding of the problems and data. Positive comments were given where some users explained how the system gives them the opportunity to see connections from a wider perspective than was previously possible. This also leads to a better handling of cases.

A positive relationship between User Satisfaction and Individual Impact supports hypothesis H7, which according to Iivari (2005) we can interpret as users' requirements are met by the systems' capabilities, and as such there exists this relationship with individual impact.

Conclusion

In this study, the Delone and McLean (2003) IS success model has been tested in a public organization context. Overall, this study supported the model, where out of the eight relationships tested, five of them were significant. User satisfaction is predicted by information quality and system quality. Also, system quality is positively associated with use and use and user satisfaction positively influence individual impact.

However, information quality is not a predictor of use and also, use does not influence user satisfaction and vice versa. We conclude that use is not fully explained in this model. We believe that other predictor variables like position within the company, BI access level, knowledge of the legacy system, understanding of the meaning of the data and sufficient information about the benefits of the system are necessary to be included during a BI system evaluation within a public organization, in this case a police organization.

References

Bailey, James E., and Pearson, S. W. 1983. "Development of a Tool for Measuring and Analyzing Computer User Satisfaction." *Management Science* 29 (5):530-545.

Bernstein, Jum C. and Nunnally, Ira H. 1994. *Psychometric theory*. 3rd ed. New York: McGraw-Hill.

Carlsson, Sven A., Skog, Linda-Marie, and Tona, Olgerta. 2010. "An IS Success Evaluation of a DSS in a Police Organization." In *Bridging the Socio-technical Gap in Decision Support Systems*, edited by Ana Resplcio, Frédéric Adam, Gloria Phillips-Wren, Carlos Teixeira and João Telhada, 443-454. Amsterdam: IOS Press.

Chin, Wynne W. 1998. "The partial least squares approach to structural equation modeling." In *Modern Methods for Business Research*, edited by G. A. Marcoulides, 295-336. Mahwah, NJ: Lawrence Erlbaum Associates.

Chin, Wynne W., and Newsted, Peter R. 1999. "Structural equaltion modeling Analysis with small samples using partial least squares." In *Statistical Strategies for Small Sample Research*, edited by Rick H. Hoyle, 307-341. Thousand Oaks, California: Sage Publications.

Cohen, Jacob. 1988. *Statistical Power Analysis for the Behavioral Sciences* 2nd ed. Hillsdale, NJ: Lawrence Erlbaum Associates, publishers.

Davis, Lois M. and Jackson, Brian A. 2005. "Acquiring, Implementing, and evaluating information technology." In *Information technology and the criminal justice system*, edited by April Pattavina, 29-57. Thousand Oakes, California: Sage Publications, Inc.

DeLone, William H., and McLean, E. R. 2003. "The DeLone and McLean model of information systems success: A ten-year update." *Journal of Management Information Systems* 19 (4):9-30.

DeLone, William H., and Ephraim R. McLean. 1992. "Information System Success: The Quest for The Dependent Variable." *Information Systems Research* no. 3 (1):60-95.

Doll, William J., and Torkzadeh, Gholamreza. 1988. "The Measurement of End User Computing Satisfaction." *MIS Quarterly* 12 (2):259-274.

Fornell, Claes R., and Larcker, David F. 1981. "Evaluating Structural Equation Models with Unobservable Variables and Measurement Error." *Journal of Marketing Research* 18 (1):39-50.

Garner, Inc. 2009. Gartner Says IT Spending to Rebound in 2010 with 3.3 Percent Growth After Worst Year Ever in 2009. Analysts Discuss the

Future of IT during Gartner Symposium/ITxpo 2009, October 18-22, in Orlando. Garner, Inc.

—. 2011. Gartner Says Worldwide Business Intelligence, Analytics and Performance Management Software Market Surpassed the $10 Billion Mark in 2010. Stamford, Conn.: Gartner Inc.

Green, Samuel B. 1991. "How many subjects does it take to do a regression analysis?" *Multivariate Behavioral Research* 26 (3):499-510.

Hair, Joe F., Ringle, Christian M., and Sarstedt, Marko. 2011. "PLS-SEM: Indeed a silver bullet." *Journal of Marketing Theory and Practice* 19 (2):139-151.

Hawking, P. and Sellitto, C. 2010. Business intelligence (BI) critical success factors. Paper read at the Australian Conference on Information Systems, at Brisbane, Australia.

Livari, Juhani. 2005. "An Empirical Test of the DeLone-McLean Model of Information System Success." *The DATA BASE for Advances in Information Systems* 36 (2):8-27.

Luftman, J., and Ben-Zvi, T. 2011. "Key issues for IT executives 2010: Cautious optimism in uncertain economic times." *MIS Quarterly Executive* 10 (4): 203-212.

McGill, Tanya J., Hobbs, Valerie J., and Klobas, Jane E. 2003. "User developed applications and information systems success: A test of DeLone and McLean's model." *Information Resources Management Journal* 16 (1):24-45.

Pick, A. R. 2008. "Benefits of Decision Support Systems." In *Handbook on Decision Support Systems 1*, edited by Frada Burstein and Clyde W. Holsapple, 719-730. Berlin: Springer.

Rai, Arun, Lang, Sandra S., and Welker, Robert B. 2002. "Assessing the Validity of IS Success Models: An Empirical Test and Theoretical Analysis." *Information Systems Research* 13 (1):50-69.

Roldán, José L., and Leal, Antonio. 2003. "A Validation Test of an Adaptation of the DeLone and McLean's Model in the Spanish EIS Field." In *Critical Reflections on Information Systems: A Systemic Approach*, edited by Jeimy J. Cano, 66-84 Hershey, PA: IDEA Group Publishing.

Rosacker, Kirsten M. and Olson, David L. 2008. "An empirical assessment of IT project selection and evaluation methods in state government." *Project Management Journal* 39 (1): 49-58.

Seddon, Peter B. 1997. "A respecification and Extension of The DeLone and McLean Model of Is Success." *Information Systems Research* 8 (3):240-253.

Seddon, Peter B. and Kiew, M-Y. 1994. "A partial test and development of the DeLone and McLean model of IS success." In *Proceedings of the International Conference on Information Systems*, edited by J.I. DeGross, S.L. Huff and M.C. Munro, 99-100. Atlanta, GA: Associations for Information Systems.

Watson, H. J., and Wixom, B. H. 2007. "The current state of Business Intelligence." *IEEE Computer* 40 (9):96-99.

CHAPTER FOUR

MODELING KNOWLEDGE MANAGEMENT SYSTEMS FOR COMPONENT-BASED SOFTWARE ENGINEERING

MOHAMMED AMINE MOSTEFAI
AND MOHAMED AHMED-NACER

Introduction

Component-based software engineering (CBSE) is a discipline that consists in the development of systems as assemblies of parts (software components), the development of parts as reusable entities, and the maintenance and upgrading of systems by customizing and replacing such parts (Crnkovic 2003). A software component is a unit of composition with contractually specified interfaces and explicit context dependencies only. A software component can be deployed independently and is subject to composition by third parties (Szyperski 2002).

CBSE activities involve many people working in different phases and activities. The knowledge used and produced in these activities is various, growing and hardly traceable: organizations have problems keeping track of what this knowledge is, where it is, and who owns it (Lindvall and Rus 2002). In this perspective, Knowledge Management (KM) provides mechanisms to create, share, transfer, and apply knowledge.

KM relies on a technological infrastructure called Knowledge Management Systems (KMS) (Alavi et al. 1999). The design of KMS uses a set of methods similar to Software Engineering Techniques. These methods are called Knowledge Engineering Techniques. Our work addresses modeling KMS for the CBSE domain by using a popular Knowledge Engineering technique: CommonKADS.

This paper is structured as follows: Section II tackles the motivations of implementing KM in CBSE and quotes related work; Section III introduces knowledge engineering techniques and justifies the choice of

CommonKADS; Section IV reports how we applied CommonKADS in the context of CBSE; and, finally, the last section summarizes our findings.

Motivations and Related Work

The main purpose of CBSE is to reduce time-to-market delays by using (or reusing) components. However, there are many risks in CBSE (Crnkovic 2003, Voas 1998): 1) an increased time and effort required for the development of components, 2) the consequences of bad component selection, 3) the dependence to the component supplier, and 4) the costs of the maintenance of components and sensitivity to their changes.

These risks can be avoided, or at least reduced, if the knowledge existing in the brain of collaborators, in the corporate memory, or outside the organization is effectively captured and used.

Many factors motivate the implementation of KM in CBSE: The need to capture and use product and process knowledge, the need to identify knowledge holders, the necessity to apply acquired knowledge in CBSE activities, the need to store and manage knowledge related to components, and the need to promote the collaboration between software engineers who are involved in CBSE activities.

The implementation of KM is not intended to be a replacement for the current software processes and tools in the organization. It is a support tool that targets making knowledge more available and effectively shared while executing these processes.

One of the most famous achievements in implementing KM in software engineering (SE) is the Experience Factory (EF) described by the Basili team (Basili et al. 1994). The EF is a technical and social infrastructure that separates the development team into two subdivisions, one dedicated to the product development and the other dedicated to the collection and the sharing of reusable experiences.

The experience factory is based on a proprietary methodology called Quality Improvement Paradigm (QIM) and addresses software engineering in general and not specifically CBSE.

Rus and Lindvall (2002) emphasize the motivations for implementing KM in SE but do not propose a model for implementation. On the other hand, Kneuper (2002) proposes an approach for supporting software processes using KM. Again, the solution is not based on a standard methodology, and there is no particular focus on CBSE.

Designing the Knowledge Management System

The implementation of KM in the organization is achieved by a special class of information systems called knowledge management systems (KMS). KMS are the tools for managing knowledge, helping organizations in problem-solving activities, and facilitating decision making (Speel et al. 2001).

KMS inherit characteristics from Knowledge-Based systems (KBS). KBS are designed using knowledge engineering techniques that are similar to software engineering techniques but with an emphasis on knowledge rather than data processing (Studer et al. 1998).

Knowledge engineering techniques are based on knowledge modeling, a special form of requirement engineering. A knowledge model provides an implementation-independent specification of knowledge in an application domain (Speel et al. 2001).

The development process of a KBS is similar to the development of any other software system. Phases such as requirements elicitation, system analysis, system design, system development, and implementation are common activities.

The stages in KBS development are business modeling, conceptual modeling, knowledge acquisition, knowledge system design, and KBS implementation.

Modeling contributes to the understanding of the source of knowledge, the flow of knowledge, and the identification of other variables such as the impact that management action has on the organizational knowledge (Davenport and Prusak 2000).

Knowledge modeling is used to provide a conceptual view of the KMS. The constructed models make the KMS more explicit and faster to develop.

There are many techniques of knowledge modeling, such as CommonKADS (Schreiber 2000), Protégé 2000 (Grosso et al. 1999) or MIKE (Angele et al. 1996).

We have chosen CommonKADS because of the pertinence of the models it includes and their large functional coverage. In addition to that, CommonKADS is one of the most-used methodologies for modeling knowledge. Moreover, the models of CommonKADS permit a very detailed view of all the aspects of the KMS by using the Unified Modeling Language (UML) notation, which is widely used in the Software Engineering community.

Protégé 2000 has a very intuitive user interface and provides a very strong way to model knowledge but it is not really a knowledge engineering methodology. It focuses rather on creating and editing ontologies. MIKE, by contrast, is suitable to KMS in a development state but does not support standards such as UML.

The CommonKADS has evolved from the KADS to the formulation proposed by Schreiber (2000). The CommonKADS methodology provides an integrated framework for knowledge management, which also encapsulates knowledge engineering activities.

CommonKADS supports structured knowledge engineering techniques, provides tools for corporate knowledge management, and includes methods that perform a detailed analysis of knowledge-intensive tasks and processes.

CommonKADS embraces an object-oriented development process and uses UML diagrams, such as class diagrams, use-case diagrams, activity diagrams and state diagrams. CommonKADS also has its own graphical notations for task decomposition, inference structures and domain schema generation. CommonKADS addresses the complexity of the system by using different models (Schreiber 2000, Motta 2001).

The organizational model provides a view of the organization where the KMS will be deployed. It also addresses the impact of the KMS on this organization.

The task model refers to the characteristics of the business processes. The characteristics are the inputs and outputs, the preconditions, performance and quality, the function of the agents that will carry out the processing, the structural coupling of these agents, the flow of knowledge between the agents, their overall control, the knowledge and skills of the agents, and the resources available to deliver the business process.

While the task model specifies what needs to be done, the agent model specifies who performs it. The task model describes "the characteristics of agents, their competencies, authority to act, and constraints in this respect." The term "agent" in CommonKADS has a generic connotation: an agent can be a human being, a robot, or a software program.

The knowledge model is used to describe the application-related knowledge used to perform tasks and the role of the knowledge in problem-solving activities. The knowledge model of CommonKADS has three categories of knowledge: task knowledge that describes the execution order for the reasoning steps, inference knowledge that describes the reasoning step performed using the domain knowledge and the domain knowledge that describes the application concepts.

The communication model describes the inter-agent communication needed when performing the tasks. The design model specifies the target software and hardware platform, the various software modules included in the target system, their functional and technical specifications, and the mapping between these modules and the conceptual components identified during the analysis phase.

In CommonKADS, designing the KMS entails the construction of the models listed above. The models are structured in three levels: the context level, the concept level, and the artefact level. Each level is composed of one or many models, and each model needs the fulfilling of the corresponding worksheets.

The models of the context level are the input for the concept level models that are, in turn, the input to the final model (the design model).

Figure 1: The CommonKADS Models

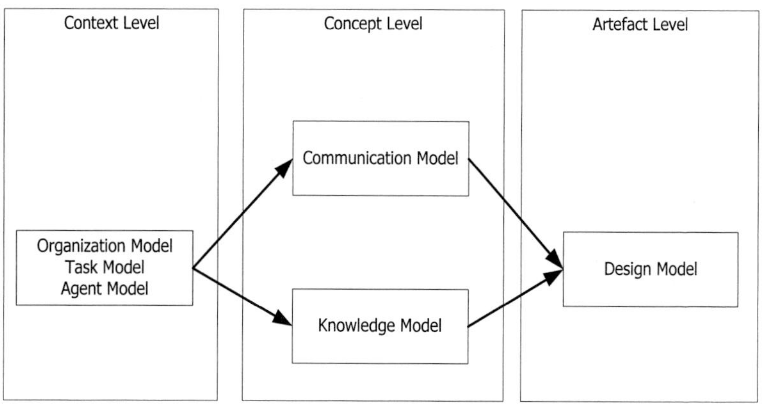

The context level includes the organization model, the task model, and the agent model. The organization model (OM) focuses on the organization structure, the processes, the staff, and resources (AIFB 2001). The OM is composed of five worksheets: OM-1 (problems, opportunities, and solutions), OM-2 (structures and processes), OM-3 (detailed tasks), OM-4 (knowledge assets), and OM-5 (feasibility). The task model focuses on the processes executed in the organization and is composed of two worksheets: TM-1 (Tasks details) and TM-2 (Tasks knowledge). The agent model deals with the profiles in the organization and their missions. This model includes one worksheet, AM-1 (Agent description). An additional worksheet, OTA-1 (impacts and improvement decisions), summarizes this level.

The concept level includes the communication model (worksheets CM-1 and CM-2) and the knowledge model. The artefact level is composed of the design model that represents the technological infrastructure of the KMS.

Modeling Knowledge Management Systems for CBSE

Modeling Considerations

The models of CommonKADS are designed with the vision of being independent from organization-specific aspects such as tools, methods or culture.

In the implementation phase, each organization has to adapt the models to its own context by extending the model and/or including organization-specific considerations in the implementation phase.

We apply the CommonKADS to construct the models level by level, and, consequently, in three phases. In the first phase, we build the three context level models: the organization model, the task model and the agent model. In the second phase, we build the knowledge model and the communication model. Finally, the models built in the previous phases will be the input for the design model.

Phase 1: The Context Level

OM-1, Problems and Opportunities

This worksheet focuses on the essential problems or opportunities concerning CBSE in an organization, attempts to contextualize these problems and provides some propositions for solutions. The main targeted problems are the motivations of implementing KM in SE, discussed in section II. The problems are divided into two categories. The first category concerns general software engineering problems, summarized in Rus and Lindvall (2002), such as capturing processes knowledge, making employees more efficient and indirectly, enhancing quality and reducing time to market. The second category concerns CBSE-specific problems, outlined by Crnkovic (2003), such as the impacts of bad component selection, insufficient analysis/design, architecture mismatch, or the costs of development and maintenance.

All the problems related to both categories have formal and informal methodologies. For example, all formal and informal component evaluation methodologies cited in Goulão and Abreu (2004) are intensively based on knowledge. An efficient management of this knowledge will be an excellent support to these methodologies.

Knowledge management does not intend to be the direct solution for all the problems. KM is a supporting solution that enhances the capture and the access to knowledge that is very precious to the problem-solving activities related to CBSE.

OM-2, Structure and Processes

In this worksheet we focus on the organization structures, involved people, the corresponding processes, the utilized knowledge, and the related culture.

The main CBSE structures can be projected on CBSE activities: each activity will be a logical structure. The CBSE activities, by inheritance to standard SE, can be structured in two categories: product-oriented activities, and process-oriented activities (Rombach and Verlage 1995).

The process-oriented activities include improvement and modeling of processes. The product-oriented activities are subdivided in two categories: managerial activities (product management, project management, project data management …), and technical activities (requirements, design, implementation, integration, component selection, component adaptation, component integration, test, release, maintenance …) (Chaudron and Crnkovic 2007).

Each activity determines the people involved, the resources used, and the knowledge manipulated. For example, test activities implicate testers, quality insurance tools, automated test tools, test-oriented knowledge, etc. Implementation involves coders, integrated development environments, compilers, etc. Component selection involves component selectors, developers, project manager, component database, internet, component-oriented knowledge, etc.

Each process will be described using UML notation. For example, the activity diagram provided in the example (Figure 2) illustrates a typical component-testing activity involving two user profiles (tester, developer) and various knowledge assets (requirements, test plans …).

OM-3, Process Breakdown

In this phase, the processes listed in OM-2 are detailed using the OM-3 worksheet provided by CommonKADS. Every single process is assigned a unique identifier and a name. For each task, we indicate 1) who performs it, 2) where it is performed, 3) the knowledge it requires, 4) if it is knowledge-intensive, and 5) its significance.

The table below (Table 1), provides an example of process breakdown. The processes we took in the examples are a subset of CBSE activities based on the general process model for CBSE provided by Dogru and Tanik (2003) and Chaudron and Crnkovic (2007).

Figure 2: Example of an Activity Diagram Illustrating the Component Testing Process

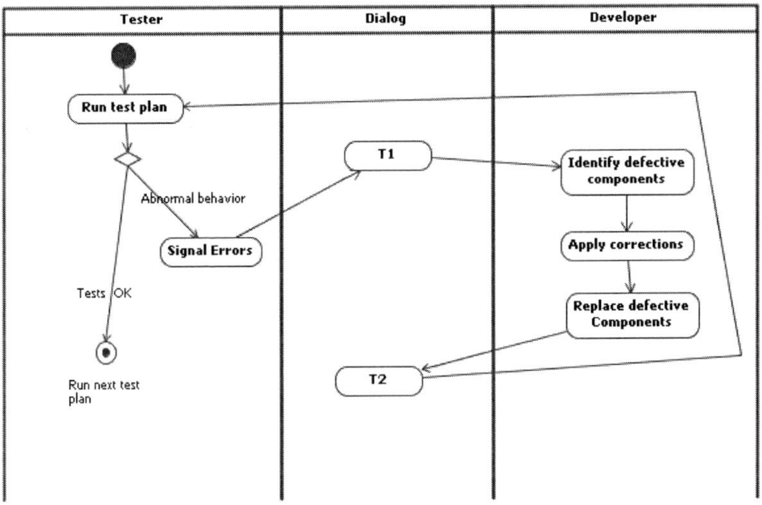

All the tasks listed in Table 1 are knowledge-intensive and make use of two types of knowledge: tacit knowledge (individual and informal knowledge stored in an individual's brain, e.g., individual knowledge and experience), and explicit knowledge (explicitly formalized and generally stored in an electronic format, e.g., documents, databases, manuals, internet) (Nonaka and Konno 1998).

For example, for administrative and technical reasons, the component search can be done by managers, architects, and developers. This task is achieved in the corresponding structures by using both kinds of knowledge.

Table 1: Process Breakdown

ORGANIZATIONAL MODEL		OM-3, PROCESS BREAKDOWN				
TASK NO	TASK	PERFORMED BY	WHERE?	KNOWELEDGE ASSET	INTENSIVE?	SIGNIFICANCE
1	SOFTWARE SYSTEM SPECIFICATION	P. MANAGER ARCHITECT	MANAGEMENT ARCHITECTURE	Specification documents, individual experience	YES	
2	DECOMPOSITION	P.MANAGER ARCHITECT	MANAGEMENT ARCHITECTURE	Individual experience, previous projects packages, components interfaces	YES	
3	COMPONENT SPECIFICATION	P.MANAGER ARCHITECT	MANAGEMENT ARCHITECTURE	Individual experience, templates	YES	

4	COMPONENT SEARCH	P.MANAGER ARCHITECT DEVELOPER	MANAGEMENT ARCHITECTURE DEV WORKSHOPS	Component database, internet, individual experience	YES	
5	COMPONENT MODIFICATION	DEVELOPER	DEV WORKSHOPS	Component documents and manuals, individual experience	YES	
6	COMPONENT CREATION	DEVELOPER	DEV WORKSHOPS	Developer knowledge, manuals, trainings, documents, Source codes	YES	
7	INTEGRATION	DEVELOPER	DEV WORKSHOPS	Developer knowledge, reports, components database, internet	YES	

8	TEST	TESTER	TEST DEP	Tester knowledge, test templates, trainings, documents	YES	

OM-4, Knowledge Assets

In this step, for each knowledge asset listed in the previous step (OM-3), we specify the agent possessing the asset, the tasks in which the asset is used, and whether it is in the right form, in the right place, in the right time, and in the right quality (Schreiber 2000).

For example, the generic knowledge asset "individual experience" is used in all tasks. The knowledge asset "test templates" is used in the tasks associated with testing and is possessed by the user with the "tester" profile. The questions "right place?", "right time?", "right form?" and "right quality?" depend on the organization and the culture of the environment.

Because of its importance and sensitivity, the challenge is to transform the tacit knowledge (individual experience) to a more exploitable form (externalization process) (Nonaka and Konno 1998). However, in most instances it is not easy to do. The designers of the KMS and the KM project managers have to give importance to the socialization process, too (direct and informal knowledge-sharing among individuals).

OM-5, Feasibility Decisions

The "OM-1" to "OM-4" steps are concerned with specifing the context of CBSE (organization structure, agents, processes, knowledge). The OM-5 step establishes judgment on the feasibility of implementing the KMS in the context of CBSE. To study the feasibility of the KMS project, we study the factors that can hamper the project's feasibility and success. These factors are regrouped in three axes: 1) financial factors, 2) technical factors, and 3) social factors.

The most significant financial factor is that KM implementation will involve important human resources and will require more time and consequently more money. The project managers are focused on finishing current projects on time rather than on long-term additional projects.

However, even if KM implementation might be costly in the short term, it addresses risks that can be fatal to the organization, such as 1) loss of knowledge, 2) lack of knowledge, 3) repetition of mistakes, or 4) unavailability of people owning key knowledge (Rus and Lindvall 2002).

The technical factors are that 1) organizations already have a set of installed tools and PSEE (Process-centered Software Engineering Environments), so the integration of the KMS should be light and transparent, and 2) the complexity of representing and using tacit knowledge.

Fortunately, most software engineering tools and environments are built on mechanisms that allow easy integration, such as APIs or web services (Wicks 2006).

Moreover, the SE domain has the particularity that people are accustomed to information systems and IT tools, and that a very significant part of data is already stored in an electronic form (Schneider 2001).

The knowledge model, discussed in the next sections, will address the knowledge representation problem.

Finally, social factors such as the non-development of share culture, reluctance to share and/or receive knowledge, and fear of sharing negative experiences could make such a project a failure.

To face these issues, implementing KM does not rely only on a technological infrastructure; it is accompanied by a suitable methodology and strategy to ensure success factors. These factors include motivational practices, knowledge champions, corporate awareness campaigns, anonymous participation, etc. (Rus and Lindvall 2002, Davenport et al. 1998).

TM-1, Tasks Descriptions

In this step, each task obtained in OM-3 is analyzed from two points of view: managerial, and informational. The managerial point of view details the goal of the tasks, the added-values, and measures used to qualify the quality and the performance of the task (Schreiber 2000).

The informational point of view discusses the functional view of the task (schematized by an activity diagram), the structure of manipulated objects, their relationships (schematized by a class diagram), and the description of the temporal order of, and control over, the subtasks (schematized by a state diagram) (AIFB 2001).

Figure 3: Component Selection Objects Modeled by a UML Class Diagram

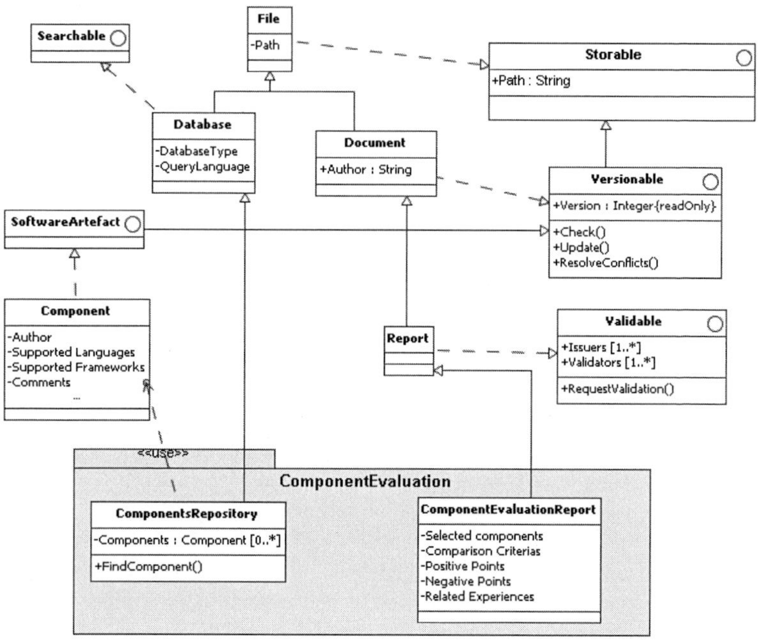

For example, Figure 3 models a subset of the "component selection" activity objects. The two main objects are the evaluation report and the component repository.

The evaluation report gleans information from earlier reports and, indirectly, from documents and files. Because the evaluation report is a report, it is a "Validable" object and, because it is a document, it implements the "Storable" and "Versionable" interfaces. On the other hand, "ComponentRepository" is a database containing a set of component versions. While the static aspect of the task is described using class diagrams, dynamic aspects are models that use activity diagrams or state diagrams.

For the example below (Figure 4), the subtask of the selection induces three states: component search, component evaluation, and component approval. In the first state, the component repository modeled previously is used. In the other states, the component evaluation report is used.

Figure 4: Component Selection Subtasks Modeled Using UML States Diagram

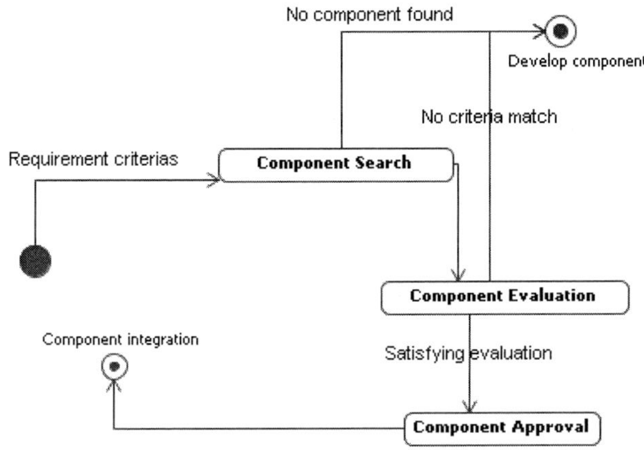

TM-2, Task Analysis

This step is a refinement of the OM-4 step, where we specify for each knowledge asset of OM-4 the following criteria: the knowledge nature (formal, rigorous, empirical, heuristic, action-based, incomplete …), the knowledge form (mind, paper, electronic…), and the knowledge availability (limitation in time, limitation in space, limitation in access, etc.). The analysis in this step cannot be generic and is toughly organization-dependent. For example, a test plan can be more formal in an organization than another depending on the adopted culture and methods.

AM-1, The Agent Model

In this step we focus on the various roles and skills in a general CBSE environment. Using the worksheet, we indicate the following for each agent: his name, his organization (basing on OM-2), the tasks he is involved in (TM-1), the agents he communicates with, knowledge he possesses (OM-4), any additional competencies, and his responsibilities and constraints.

In addition to AM-2 worksheets, UML uses case diagrams to schematize agent actions. For example, Table 2 illustrates the AM-2 worksheet for the "component selector" role.

Table 2: AM-1 Worksheet For The Role "Component Selector"

Agent criteria	Value
NAME	COMPONENT SELECTOR
ORGANIZATION	COMPONENT SELECTION COMPONENT EVALUATION DEVELOPMENT
INVOLVED IN	COMPONENT SELECTION TEST
COMMUNICATES WITH	PROJECT MANAGER DEVELOPER TESTER
KNOWLEDGE	INDIVIDUAL EXPERIENCE COMPONENT EVALUATION REPORTS
OTHER COMPETENCIES	(ORGANIZATION-DEPENDANT)
RESPONSIBILITIES AND CONSTRAINTS	- FIND AND EVALUATE COMPONENTS - THE EVALUATION QUALITY IS VITAL TO THE SUCCESS OF THE PROJECT

OTA-1, Impacts and Improvements

This is the final step of the context phase. It deals with the impacts of KMS implementation and the changes that will have to be adopted for said implementation.

In a modeling point of view, the availability, the efficiency of transfer channels, and the quality of knowledge are vital factors for improving the various CBSE tasks and activities.

If correctly implemented and accompanied by a suitable strategy, the KMS will be widely beneficial and will be a part of the solution of all knowledge-intensive tasks problems. In an implementation point of view, each organization has to adapt and/or ameliorate the model to fit its culture and integrate with its tools. In that sense, OTA-1 is widely organization-dependent.

The Concept Level

The Knowledge Model

The knowledge model of an application provides a specification of the data and knowledge structures required for the application (Speel et al. 2001).

Knowledge is structured in three categories: 1) domain knowledge including static concepts and domain types, 2) task knowledge, which is a functional decomposition of the activities and is goal-oriented, and 3) inference knowledge that includes basic inference steps (using domain concepts) to undertake an action.

The domain knowledge captures static structures and types concerning CBSE. Ontologies are used to specify this knowledge. Ontology is defined as an explicit specification of the terms and their meaning in the domain of interest. The role of ontologies is to provide a clear and consensual knowledge structure, reduce the ambiguity due to terminological conflicts, and enhance knowledge share and reuse.

Ontologies specify knowledge by using classes (concepts) and slots (properties). While ontologies handle the formal aspect of concepts as classes, knowledge-bases hold instances of these concepts as class-instances.

There have been many attempts to define ontologies related to SE (Calero et al. 2006), but in order to reach consensus, the most promising achievements are based on the project SWEBOK (IEEE 2012). The project SWEBOK aims at providing a standardized set of all SE engineering definitions and concepts.

Most large-coverage SE ontologies, such as Mendes and Abran (2004) or Sicilia et al. (2005) are based on the SWEBOK body of knowledge. For our work, we will use the ontology (and ontology development methodology) of Mendes and Abran (2004) because it provides a high level of details (more than 4000 concepts) and a clear methodology.

According to the SWEBOK project, the SE domain has been divided into ten categories representing ten knowledge areas in the ontology: requirements, design, construction, testing, maintenance, configuration management, software engineering management, software engineering processes, tools and methods, and finally, software quality (IEEE 2012).

While each knowledge area possesses its own concepts and relations, some "shared concepts" are used in more than one knowledge area. These shared concepts are derived essentially from the concepts Artefact, Activity, or Agent (Abran et al. 2004).

For example, in Figure 5, a software component is a shared concept descending from the concept "Artefact." Component selection activity is derived from the "determine deliverables" activity that belongs to "software engineering management" knowledge area.

While domain and tasks knowledge address CBSE-specific concepts, inference knowledge is about basic reasoning tasks explicitly or implicitly used in every-day activities.

Inference knowledge is represented using functions and roles. A function is a basic reasoning step included in one or many tasks. A knowledge role is knowledge used as input or output for functions (Schreiber 2000).

Figure 5: The Component Selection Related Concepts Derived from SWEBOK Concepts

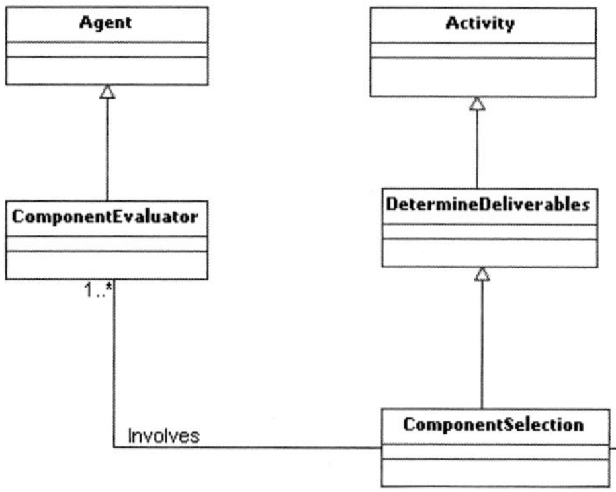

The inferences are the lower level of tasks hierarchy. Each CBSE task implies one or, more generally, a set of inferences. On the other hand, a single inference can be used by more than one task.

For the modeling of the KMS, even if the activities enumerated in (IEEE 2012) are in a high level (tasks), the modeling of inference will be achieved by decomposing these tasks and extracting basic reasoning activities as inferences.

Here again, inferences will be categorized in two groups: shared inferences used in more than a single knowledge area, and specific inferences exclusive to only one knowledge area.

For example, the “IntegrateComponent” inference (Figure 6) has, for input, the component to integrate, the module to integrate into, the integration manual, and an environment for integration (IDE).

This basic inference is used in the “Coding” activity (integrate a component), individual component test (integrate the component in test projects), or in the maintenance process (replace a component with another, more satisfactory component or another version of the same component).

Figure 6: The Inference "IntegrateComponent" Used in More than One Task

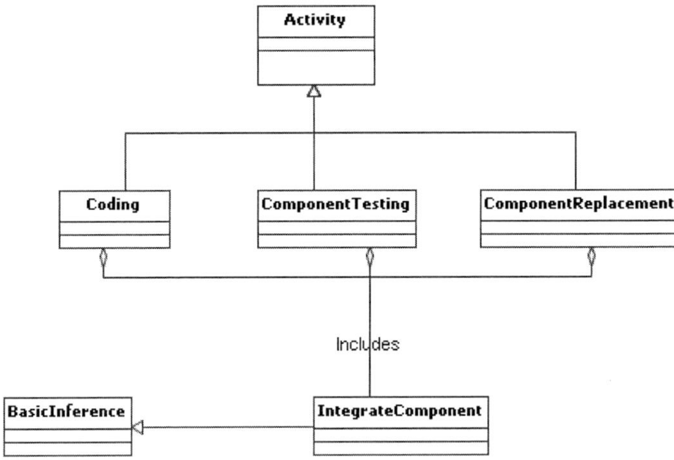

The Communication Model

The communication model provides descriptions about communications among the different agents. As such, the communication model is the model that specifies knowledge transfer among agents.

In the communication model, communications among agents are composed of communication units called "transactions". At a lower level, each transaction is composed of a set of messages.

Consequently, the process of building the communication model is structured in three layers: 1) the communication plan that governs the full dialog between two agents, 2) the individual transactions that link two (leaf) tasks carried out by two different agents, and 3) the information exchange specification that details the internal message structure of a transaction (Schreiber 2000).

In the process of constructing the communication model for CBSE, we list all the tasks (Task Model) that involve more than one agent and detail the transactions engendered by these tasks.

The CommonKADS methodology provides two worksheets, CM-1 and CM-2, allowing the construction of this model.

The CM-1 worksheet is applied to each transaction in the deducted communication plans. It specifies the transaction name, the knowledge elements included, the agents involved and the owner communication plan. The CM-2 details the messages structure of each transaction.

Messages are classified according to their types and their purpose: communication initialization messages (START, FINISH), acceptance messages (ACCEPT, REJECT), information messages (INFO, WARNING, ERROR) and request messages (REQUEST, PROPOSE). Each message indicates the agent that sent it, the reference, the contents, eventual attachment, and the receivers.

The Artefact Level: The Design Model

The design model provides a description of the software modules that will compose the KMS, describing the data structures used and preparing the implementation. The models generated by the analysis phase are the input for the design model. However, to remain context-independent, we will provide only the software architecture of the KMS. The implementation details will be processed individually by the organizations according to their platforms and tools.

The KMS Features

The KMS should ensure the following additional functions (Staniszkis 2003) 1) Contents Repository: storage of knowledge instances, 2) Knowledge Dissemination: making knowledge accessible to KMS users, 3) Content Integration: integrating contents provided by external sources (documents, files, databases, etc.), 4) Collaboration: communication mechanisms among KMS users (mails, chats, wikis, forums, etc.), 5) Security: protecting knowledge.

The KMS Functional Model

The KMS functional model (see Figure 8) provides the architectural foundation for implementing KMS in the organization.

According to the requirements expressed in (Staniszkis 2003) and the models created in the analysis phase, the modules composing the functional model are:

a) The Meta-Data Layer: The meta-data layer includes all the knowledge definitions. It contains the classes and relationships expressed in the knowledge model. To suit the organization requirements, the classes and relationships are structured as an ontology extending the ontology provided in the knowledge model.

b) The Knowledge Base: While the meta-data layer stores the definitions of classes, the knowledge-base (KB) will store all the instances of these classes called knowledge instances. The knowledge base allows the persistence of objects and provides mechanisms to query the KB. The KB module can be implemented using relational databases or object databases.

c) Content Integration Service: Ensures the automatic collection of knowledge from external knowledge sources such as files, source codes, databases, document repositories, internet, external KMS, etc.

d) Personalization Service: Adapts (form and content) the knowledge elements to the profile of the current user. For examples, this service filters information considered "too technical" for non-technical profile.

e) Security Service: Authenticates collaborators and processes, and defines their access level and action perimeter.

Figure 8: The KMS Functional Model

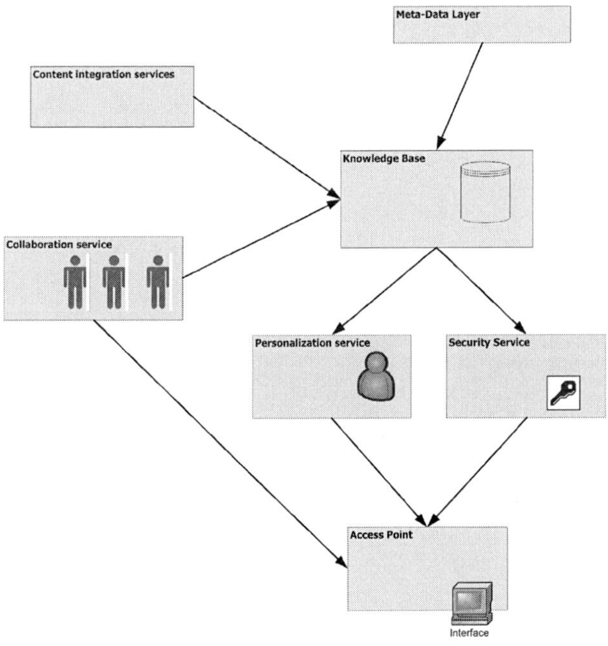

f) Collaboration Service: Provides a set of tools allowing the collaborators to be linked in (by using forums, chats, mailings, etc.). The collaboration service implements all the transactions listed in the communication model. In addition, this service provides a gateway to the content integration service for capturing knowledge created or transferred during communications.

g) The Access Point: The access point allows entering knowledge queries, creating or updating knowledge instances and evaluating existing knowledge instances.

Conclusion and Future Work

Our work was about modeling Knowledge Management Systems for Component Based Software Engineering.

Among Knowledge Engineering Techniques, we have chosen CommonKADS because the models it includes target all the aspects of the KMS from knowledge assets to functional and communicational behavior. When applying CommonKADS methodology, we have considered as a constraint the independence from organization-specific context such as tools, methods, or culture, and the consideration of integration of the KMS with the organization tools.

In an organization-independent context, we constructed the CommonKADS models: the task model, the agent model, and the communication model.

For the knowledge model, instead of developing an ontology from scratch, we adopted a consensual ontology with construction based on the SWEBOK project.

Finally, we brought a design model serving as the basis for implementation of KMS in the context of the organization.

However, this paper focuses only on the modeling aspects, as well as on some practical and technological issues that have to be addressed, in order to list guidelines for the implementation of KMS.

References

Abran, A., P. Bourque, R. Dupuis, J. Moore, and L. Tripp. 2004. *Guide to the Software Engineering Body of Knowledge – SWEBOK.* Edited by Alain Abran, Pierre Bourque, Robert Dupuis, James Moore and Leonard Tripp: IEEE Press.

AIFB. 2001. The CommonKADS Methodology. Institut AIFB Universität Karlsruhe 2001.

Alavi, M., R. H. Smith, and D. E. Leidner. 1999. "Knowledge management systems: issues, challenges, and benefits." *Communications of the AIS* no. 1 (2).

Angele, J., D. Fensel, and R. Studer. 1996. "Domain and task modelling in MIKE." In *Domain Knowledge for Interactive System Design*, edited by Chapman & Hall.

Basili, V. R., G. Caldiera, and H. D. Rombach. 1994. "The Experience Factory." *Encyclopedia of Software Engineering* no. 2.

Calero, C., F. Ruiz, and M. Piattini. 2006. *Ontologies for Software Engineering and Software Technology*: Springer.

Chaudron, M. R. V., and I. Crnkovic. 2007. Component Models and CBSE processes. eindhoven university.

Crnkovic, I. 2003. "Component-based software engineering: New challenges in software development." *CIT. Journal of computing and information technology*:151-161.

Davenport, T. H., D. W. D. Long, and M. C. Beers. 1998. "Building Successful Knowledge Management Projects." *Sloan Management Review* no. 39 (2):43-57.

Davenport, T. H., and L. Prusak. 2000. *Working Knowledge: How Organizations Manage What They Know*: Harvard Business Press.

Dogru, A. H., and M. M. Tanik. 2003. "CommonKADS Context Models." *IEEE Software* no. 34:34-41.

Goulão, M., and F. B. Abreu. 2004. Software Components Evaluation: an Overview. In *5a CAPSI*. Lisbon, Portugal.

Grosso, W. E., H. Eriksson, R. W. Fergerson, J. H. Gennari, S. W. Tu, and M. A. Musen. 1999. "Knowledge modeling at the millennium (The design and evolution of Protégé-2000)." *Proceedings of the Twelfth Workshop on Knowledge Acquisition, Modeling, and Management, Banff, Canada*.

IEEE. 2012. *Guide to the Software Engineering Body of Knowledge (SWEBOK)* 20122012. Available from http://www.computer.org/portal /web/swebok.

Kneuper, R. 2002. "Supporting software processes using knowledge management." In *Handbook of Software Engineering and Knowledge Engineering*. World Scientic Publishing Company.

Lindvall, M., and I. Rus. 2002. Lessons Learned from Implementing Experience Factories in Software Organizations. In *Fifth International Workshop on Learning Software Organizations*. Bonner Köllen Verlag, Luzern, Switzerland.

Mendes, O., and A. Abran. 2004. "Software Engineering Ontology: A Development Methodology." *Metrics News* no. 9:68-76.

Motta, E. 2001. "The Knowledge modelling paradigm in knowledge engineering." In *Handbook of software engineering and knowledge engineering: fundamentals*, edited by S. K. Chang, 589-614. Singapore: World Scientific Publishing Co.

Nonaka, I., and N. Konno. 1998. "The concept of "Ba": building a foundation for knowledge creation." *California Management Review* no. 40 (3):40-54.

Rombach, H. D., and M. Verlage. 1995. "Directions in Software Process Research." In *Advances in Computers*, edited by Zelkowitz Marvin, 1-63. Elsevier.

Rus, I., and M. Lindvall. 2002. "Guest Editors' Introduction: Knowledge Management in Software Engineering." *IEEE Softw.* no. 19 (3):26-38. doi: 10.1109/ms.2002.1003450.

Schneider, K. 2001. Experience Magnets – Attracting Experiences, Not Just Storing Them. In *Focused Software Process Improvement, PrOFES'01*. Kaiserslautern, Germany.

Schreiber, G. 2000. *Knowledge Engineering and Management: The CommonKADS Methodology*: MIT Press.

Sicilia, M.-Á., J.-J. Cuadrado, E. García, D. Rodríguez, and J. R. Hilera. 2005. The Evaluation of ontological representations of the SWEBOK as a revision tool. In *29th Annual International Computer Software and Application Conference (COMPSAC),*. Edinburgh, UK.

Speel, P. H., A. T. Schreiber, W. v. Joolingen, G. v. Heijst, and G. J. Beijer. 2001. "Conceptual models for knowledge-based systems." In *Encyclopaedia of Computer Science and Technology*, edited by A. Kent, & Williams.

Staniszkis, W. 2003. Feature Requirements of a Knowledge Management System. Rodan Systems S.A.

Studer, R., V. R. Benjamins, and D. Fensel. 1998. "Knowledge engineering: principles and methods." *Data Knowl. Eng.* no. 25 (1-2):161-197. doi: 10.1016/s0169-023x(97)00056-6.

Szyperski, C. 2002. *Component Software: Beyond Object-Oriented Programming*: Addison-Wesley Longman Publishing Co., Inc.

Voas, J. 1998. "COTS Software: The Economical Choice?" *IEEE Softw.* no. 15 (2):16-19. doi: 10.1109/52.663777.

Wicks, M. N. 2006. Tool integration in software engineering: an annotated bibliography. Heriot-Watt University.

CHAPTER FIVE

A METHOD FOR IMPROVING ALGORITHMS OF FORMAL CONCEPTS EXTRACTION USING PRIME NUMBERS

AFEF SELMI, MOHAMED MOHSEN GAMMOUDI AND FARAH HARRATHI

Introduction

Formal Concept Analysis (FCA) is still used in many fields: Information Retrieval (Carpineto and Romano 1996, Farhat et al. 2012), Machine Learning (Kuznetsov 2004), Software Engineering (Priss 2003), etc. These fields employ FCA for three main tasks: (1) to extract formal concepts (Simon 2009, Ganter 1984, Norris 1978, Chein 1969, Vilem 2008); (2) to construct the Galois lattice (Vicky 2006, Bordat 1986, Nourine and Olivier 1999, Godin 1995) or (3) to extract a concepts graph (Gammoudi 2005, Gammoudi and Nafkha 2002). The algorithms of extracting all concepts set are highly costed, considering the fact that they are NP-complete problems (Bjoem 2006). The majority of algorithms proposed in the literature tend to propose approximate solutions with a polynomial complexity (Simon 2009, Ganter 1984, Norris 1978, Krajca et al. 2010, Vilem 2008). Those algorithms are concurrent with each other, minimizing their run-times. To pursue this concurrence, we propose a new method whereby Prime Numbers are used in the encoding of objects and attributes of a formal context. This paper begins with an overview of some basic FCA notions. In section 3, extraction concepts algorithms and Prime Numbers studies are presented. In section 4, we present our approach of extraction concepts using Prime Numbers. In section 5, we show experimental results to evaluate our contribution.

Formal Concept Analysis

Formal Concept Analysis is a set of techniques that allow structuring and analyzing data. It represents a duality called the Galois connection which expresses the linkage between two kinds of items: objects and attributes. In the following, we introduce some basic notions related to FCA: formal context, Galois connection, formal concept, partial order relation and Galois lattice (Ganter and Wille 1999).

A formal context is a triple (O, A, R), where O is a finite set of objects, A is a finite set of attributes and $R \subseteq O \times A$.

For $X_i \subset O$ and $Y_i \subset A$, the two Galois connection operators f and g are defined as follows: $f(X_i) = \{y \in A|\ (x, y) \in R, \forall\ x \in X_i\}$ and

$$g(Yi) = \{x \in O|\ (x, y) \in R, \forall\ y \in Y_i\}.$$

In a formal context (O, A, R), a couple (A, B), with $A \subseteq O$ and $B \subseteq A$, is called a formal concept if and only if f(A) = B and g(B) = A. For a concept C = (A, B), the set of objects A is called extension of C, denoted A = ext(C), and the set of attributes B is called intention of C, denoted B = int(C).

Let (A1, B1) and (A2, B2) be two concepts extracted from the relation R, we call << a partial order on concepts if and only if (A1, B1) << (A2, B2), $A1 \subseteq A2$ and $B2 \subseteq B1$.

Let R be a binary relation, T is the set of concepts extracted from this relation. The Galois lattice associated with R is the set of formal concepts provided with the order relation <<, denoted (T, <<).

Related Works

In this section, we present a list of algorithms allowing formal concepts extraction, describing the complexity of each one.

Algorithms of Formal Concepts Extraction

In Chein (1969), a non-incremental algorithm is proposed. This algorithm uses the method of Divide and Conquer. In fact, it divides the set of all concepts into several subsets in order to minimize the search space. The mentioned algorithm has a theoretical complexity in the order of $O(|O^3||A||L|)$ where |O| is the number of objects, |A| is the number of attributes and |L| is the number of concepts extracted from the relation R. This algorithm tends to show a good performance with small contexts.

The algorithm proposed by Norris (1978) is an incremental one. It considers the binary relation as a binary array; it looks over it line by line to extract the formal concepts. The algorithm complexity in terms of time is $O(|O^2||A||L|)$.

The algorithm proposed in Ganter (1984) is a non-incremental algorithm. Since this algorithm computes closure for only some subsets of objects, it requires little storage space. However, its performance decreases when formal contexts become denser. The algorithm complexity in terms of time is $O(|O^2||A||L|)$.

The Close by One algorithm (CbO) mentioned in Kuznetsov (1993) uses a similar notion of canonicity. In order to extract the formal concepts from a formal context, the algorithm CbO uses an intermediate structure by improving the calculation of formal concepts. It is a non-incremental algorithm and it is used with a dense formal context. Its time complexity is $O(|O^2||A||L|)$.

An incremental algorithm called Faster was proposed by Vicky (2006) which can generate formal concepts and construct the Galois lattice from a formal context using the Depth First Search (DFS) order. By increasing the size and density of the formal context, the faster algorithm tends to have a good performance in time (Vicky 2006). In a theoretical way, the time complexity of each concept c of this algorithm is in order of $(\sum_{a\in ext(c)} |cnbr(a)|)$ where cnbr(a) is the list of attributes relative to each object a.

The algorithm proposed in Vilem (2008) is incremental. Using a Depth First Search (DFS), this algorithm can determinate the intent of each concept by computing the closure of the concept already generated and the concept to be generated. The time complexity of the algorithm is $O(|O||A^2||L|)$.

Another incremental algorithm called In-Close was proposed by Simon (2009) which is based on the CbO algorithm (Kuznetsov 1993). It generates formal concepts from a formal context using the Depth First Search (DFS) order. In a theoretical way, the time complexity of In-Close algorithm is $O(|O||A^3||L|)$.

The PCbO algorithm proposed by Krajca et al. (2010) is a parallel algorithm. It can simultaneously generate disjointed sets of concepts. This algorithm is an improvement to the CbO algorithm by adding a test of canonicity which reduces the number of concepts generated by several times. In the worst case, it generates formal concepts in the same order as the CbO algorithm. The time complexity of the PCbO algorithm is $O(|O||A^3||L|)$.

In the previous section, we have presented algorithms to extract formal concepts. Most of these algorithms for extracting formal concepts are based on the comparison of strings using the operations of intersection and union, reflecting a relatively high cost of run-time. To overcome this limitation, we propose to encode objects and attributes of the formal context by prime numbers. This choice is enforced by the fact that the usage of prime numbers makes the comparison operation easier. The use of prime number encoding is state of the art. In the following, we present a brief study of the use of this approach in different areas.

Research Studies Using Prime Numbers

Multimedia: In the multimedia field, especially in databases image storage, Chang et al. (2001) have proposed a new strategy for indexing images called Generalized Prime-Number-based Matrix (GPN Matrix). This technique indexes the objects in an image and represents each spatial relationship between two objects as the product of primes. Using the GPN Matrix strategy, the storage of images in the database is reduced and the processing time of applications has been improved.

XML databases: With prime numbers, Cheng and Radha (2006) have proposed a new query optimization technique for XML databases. This technique, with the bottom-up approach, associates a prime number with each leaf of the XML tree. The parent node is labeled by the product of prime numbers associated with each label of its sons. However, with this approach, we can reach an important number assigned to the root node of a huge tree. Plus, this technique can be used with nodes having only one son. With the top-down approach, the proposed technique is to associate with the root 1 and the prime number of each node which is not a leaf a unique prime number. Each node is labeled by the product of prime numbers associated with labels of its partners and father. This optimization technique is effective when considering a dynamic update for an environment.

The same drawback of the bottom-up approach is imposed where leaves are labeled by a large number. Therefore, the storage size of the labels will have an impact on the performance of XML query processing.

Data Mining: In this field, extraction of all frequent items will lead to a large itemsets. To solve this problem, Mohammad et al. (2008) have proposed a new approach to extract the maximal frequent itemset. This approach proposes a technique for coding database storages. With the use of prime numbers, the file encoding becomes smaller and more manageable in the memory than the original database and according to the author, the database size was reduced by more than half. For the mining of the frequent sequences of items, Karam et al. (2008) have proposed an algorithm, called PRISM (PRIme encoding based Sequence Mining) which encodes each sequence block by a prime number. It uses the join operations on the blocks to determine the frequency of each candidate. According to the author, PRISM algorithm is one of the most efficient algorithms for the extraction of data sequences and is the fastest working with large data sets.

In the next section, we will present our method where we use prime numbers for the encoding of objects and attributes of the formal context in order to improve the performance of algorithms for generating formal concepts.

PrimeFaster Algorithm

In this section, we propose PrimeFaster algorithm which is our extension of Faster algorithm (Vicky 2006). This extension is based on the use of the prime numbers in the encoding of objects and attributes of the formal context.

In order to describe PrimeFaster algorithm, we begin this section with an outline of Faster algorithm (Vicky 2006), which has the following pseudo-code:

Faster algorithm

```
Compute the top concept C= (O, attr(O));
Initialize a queue Q= {C};
Compute Child(C);
while Q is not empty do
    C= dequeue(Q);
    Let X= int(C) and suppose AttrChild(X) = < S1, S2,..., Sk >;
    for i=1 to k do
        if XSi is closed then
            Denote the concept (obj(XSi), XSi) by k;
            if k does not exist then
                Compute Child(k);
                Enqueue k to Q;
            end if
            Identify k as a successor of C;
        end if
    end for
end while
```

This algorithm is explained in the work of Choi (2006).

Principle of Primefaster Algorithm

PrimeFaster algorithm consists in extracting the formal concepts from a formal context. In a first step, we assign to each object and each attribute a Prime Number, starting with the first Prime Number 2. In a second step, we do the same treatment in the algorithm presented previously in Faster algorithm. However, in this treatment we use new operators to calculate the intersection and the union. These operators are used, respectively to compute the child of concept and to identify the successor of concept. In the following subsection we define these operators.

Intersection Operator

Let A and B be two sets, the intersection between A and B noted A ∩ B, is equal to all elements of A belonging to B.

Formally A ∩ B = {x ∈ A and x ∈ B}.

For example if A = {a,b,c,f} and B = {a,b,e,f} then A ∩ B = {a,b,f}

By encoding elements of A and elements of B by prime numbers A and B will be respectively A = {2,3,5,7} and B = {2,3,11,7}. The intersection between A and B is obtained by two steps: first, the GCD (Greatest Common Divisor) of product of elements of A and product of elements of B is calculated. Then, the GCD is decomposed into prime factors. Formally, the intersection operator is represented by the following formulae:

$$\mathrm{A} \cap \mathrm{B} = \mathrm{Decomposition}(\mathrm{GCD}(\prod_{x \in A} x, \prod_{x \in B} x))$$

In the previous example, the intersection of A and B is calculated as below:

$$\prod_{x \in A} x = 2 * 3 * 5 * 7 = 210$$

$$\prod_{x \in B} x = 2 * 3 * 11 * 7 = 462$$

$$\mathrm{GCD}(\prod_{x \in A} x, \prod_{x \in B} x) = 42$$

$$\mathrm{A} \cap \mathrm{B} = \mathrm{Decomposition}(42) = \{2,3,7\}$$

Union Operator

Let A and B be two sets, the union between A and B noted A ∪ B, is equal to all elements of A and elements of B.

Formally A ∪ B = {x ∈ A or x ∈ B}.

For example if A = {a,b,c,f} and B = {a,b,e,f} then A ∪ B = {a,b,c,f,e}

By encoding elements of A and elements of B by prime numbers, A and B will be respectively A = {2,3,5,7} and B = {2,3,11,7}. The union between A and B is obtained by two steps: first, the LCM (Least Common Divisor) of product of elements of A and product of elements of B is calculated. Then, the LCM is decomposed into prime factors. Formally, the union operator is represented by the following formulae:

$$\mathrm{A} \cup \mathrm{B} = \mathrm{Decomposition}(\mathrm{LCM}(\prod_{x \in A} x, \prod_{x \in B} x))$$

$$\mathrm{LCM}(\prod_{x \in A} x, \prod_{x \in B} x) = \frac{\prod_{x \in A} x \,*\, \prod_{x \in B} x}{\mathrm{GCD}(\prod_{x \in A} x, \prod_{x \in B} x)}$$

In the previous example, the union of A and B is calculated as below:

$$\prod_{x \in A} x = 2 * 3 * 5 * 7 = 210$$

$$\prod_{x \in B} x = 2 * 3 * 11 * 7 = 462$$

$$\prod_{x \in A} x * \prod_{x \in B} x = 210 * 462 = 97020$$

$$\mathrm{GCD}(\prod_{x \in A} x, \prod_{x \in B} x) = 42$$

$$\mathrm{LCM}(210{,}462) = \frac{97020}{42} = 2310$$

$$\mathrm{A} \cup \mathrm{B} = \mathrm{Decomposition}(2310) = \{2{,}3{,}5{,}7{,}11\}$$

In order to clarify the principle of PrimeFaster algorithm, we propose the following example.

Illustration: Let the set of objects O = {1,2,3,4,5} and the set of attributes A = {a,b,c,d}.

Table 1: A formal context (O,A,R)

R	a	b	C	d
1	1	0	0	1
2	0	1	1	1
3	0	0	1	0
4	0	1	0	1
5	1	1	1	1

Table 2: A formal context encoded by prime numbers

R	2	3	5	7	image
2	1	0	0	1	14
3	0	1	1	1	105
5	0	0	1	0	5
7	0	1	0	1	21
11	1	1	1	1	210
antecedent	22	231	165	462	

We calculate the image for each object i ∈ O and the antecedent for each attribute i ∈ A (for example, image(2) = 2 × 7 = 14 et antecedent(2) = 2 × 11 = 22).

Let the top concept C = (obj(X), X) = (2×3×5×7×11, 1) = (2310, 1), (we attribute 1 for the intention X). In order to compute Child(C), we should compute res(X) and AttrChild(X) defined in (Vilem 2008).

Let O = {2,3,5,7,11} and A = {2,3,5,7}.

(a) Compute res(X) by determining obj(X) ∩ antecedent(i) where i ∈ A \ X

i = 2, {2,3,5,7,11} ∩ {2,11} = decomposition(GCD(2310,22)) = {2,11}

i = 3, {2,3,5,7,11} ∩ {3,7,11} = decomposition(GCD(2310,231)) = {3,7,11}

i = 5, {2,3,5,7,11} ∩ {3,5,11} = decomposition(GCD(2310,165)) = {3,5,11}

i = 7, {2,3,5,7,11} ∩ {2,3,7,11} = decomposition(GCD(2310,462)) = {2,3,7,11}

res(X) = {2,3,5,7}

(a) Compute AttrChild(X). For each i ∈ res(X), we compute antecedent(i)

AttrChild(X) = {(22,2), (231,3), (165,5), (462,7)}

(b) Compute Child(C). For each intention i ∈ AttrChild(X),
we test if i is closed.
i=2, attr(obj(2)) = attr(2,11) = (2,7) ≠ 2, 2 is not closed.
i=3, attr(obj(3)) = attr(3,7,11) = (3,7) ≠ 3, 3 is not closed.
i=5, attr(obj(5)) = attr(3,5,11) = 5, 5 is closed.
i=7, attr(obj(7)) = attr(2,3,7,11) = 7, 7 is closed.
Child(C) = {(3.5.11,5), (2.3.7.11,7)}

(c) Compute Succ(C), we use top concept C = (obj(X), X) *= (2310, 1) and Child(C)*

Succ(C)={({2,3,5,7,11} ∩ {3,5,11}, {1} ∪ {5});

({2,3,5,7,11}∩{2,3,7,11},{1}∪{7})}={(decomposition(GCD(2310,165)),decomposition(LCM(1,5)));(decomposition(GCD(2310,462)),decomposition(LCM(1,7)))} ={(3.5.11,5); (2.3.7.11,7)}

For this iteration, formal concepts are (3.5.11,5) ; (2.3.7.11,7) which correspond to concepts (235,c) ; (1245,d).

Calculating Complexity

In this section, we study the complexity of the operations of intersection and union by comparing it with the operators used in Faster algorithm (Choi 2006).

Complexity of the Intersection Operator

For Faster algorithm, each intersection operation costs $O(|O^2|)$, where $|O|$ is the number of objects. In our algorithm, we used the GCD to compute the intersection of the extensions. For this aim, we performed multiplication operations where each one is costing $O(|O|)$. We know that the complexity of GCD(a,b) cost is O(log(sup(a,b))). In our case, the complexity of GCD(|O|,|O|) costs log(|O|), where |O| is the number of objects. Thus, we obtain a complexity in order $O(|O|+\log(|O|))$ which is equal to $O(|O|)$.

Complexity of the Union Operator

In Choi (2006), each union operation costs O(|A|), where |A| is the number of attributes. With our algorithm, we used the LCM to compute the union between two sets. To determine the LCM, we use multiplication, division and GCD; therefore, a union operation costs O(|A|+1+log(|A|)), which is equal to O(|A|).

In the next section, we evaluate experimentally our PrimeFaster algorithm by comparing it with the Faster algorithm (Choi 2006). For tests, we use randomly generated formal context through the Galicia[1] tool.

Experimental Evaluation

In experimentations, we use a computer with 2 Go RAM, 150 Go disk storage, and Windows 7 operating system. The algorithms Faster and our algorithm were implemented in JAVA. In order to perform our experiments, we were inspired by the work of Kuznetsov and Obiedkov (2002), to identify the evaluation parameters. These parameters are: i) the number of objects, ii) the number of attributes and iii) the density of the formal context (denoted d), which is defined by the following formulae:

$$\mathbf{d} = \frac{\boldsymbol{number\ of\ 1}}{\boldsymbol{number\ of\ objects \times number\ of\ attributes}} \times \mathbf{100}$$

We present, in the following figures, a series of tests of our algorithm as compared with the Faster algorithm (Choi 2006). We note that the run-time is in milliseconds.

Test 1: |A| = 100 and d = 25%. In this test, we set the number of attributes to 100 and the density to 25% by varying each time the number of objects.

With a formal context density equal to 25% and the number of attributes equal to 100, we observe that the PrimeFaster algorithm has a shorter run-time compared to the Faster algorithm.

[1] Galicia platform is free software; it is used to build, from a formal context, a concept lattice. http://www.iro.umontreal.ca/ galicia/download.html.

Figure 1: Diagram graph |A| = 100, d = 25%

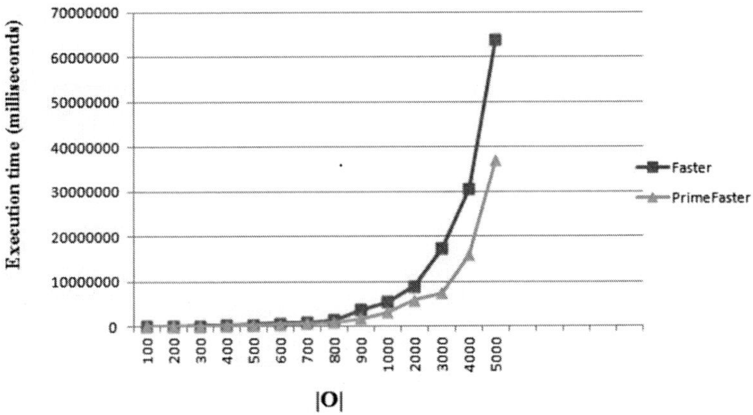

Test 2: |O| = 500 and d = 25%. For this test, we assign the number of objects to 500, the formal context density to 25%, and we vary the number of attributes.

Figure 2: Diagram graph |O| = 500, d = 25%

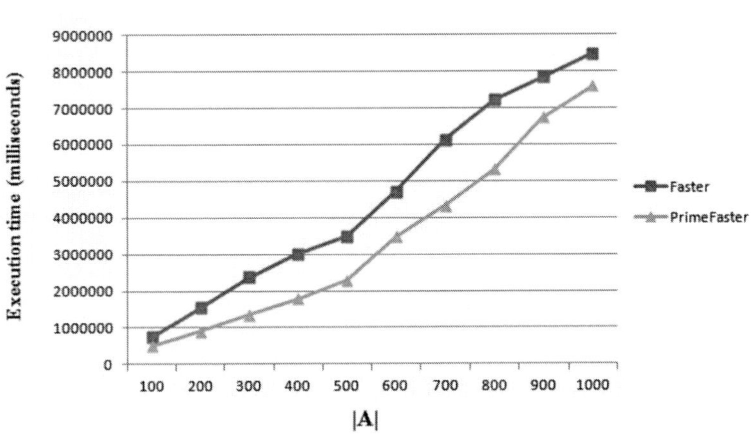

We observe as is shown in Figure 2 that even by setting the number of objects while varying the number of attributes, the run-time of the

PrimeFaster algorithm is always shorter compared to that of the Faster algorithm.

Test 3: |O| = 500 and d = 50%. In this experiment, we assign the formal context density to 50%, the number of objects to 500, and we vary the number of attributes.

Figure 3: Diagram graph |O| = 500, d = 50%

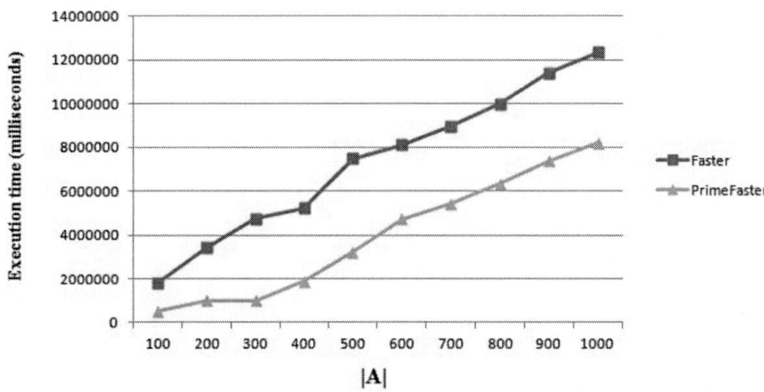

We observe, as shown in Figure 3, that the run-time of the PrimeFaster algorithm is shorter than that of the Faster algorithm. Furthermore, every time we increase the density of formal context, the difference between the two curves becomes larger.

Test 4: |O| = 500 and d = 75%. In this test, we assign the formal context density to 75%, the number of objects to 500 and we vary the number of attributes.

We obtain, as shown in Figure 4, that every time we increase the density of the formal context, the difference between the two curves increases. Indeed, the run-time of the PrimeFaster algorithm is always shorter than that of the Faster algorithm. This is due to the increased number of concepts generated from the formal context and the number of calculated intersections.

Figure 4: Diagram graph |O| = 500, d = 75%

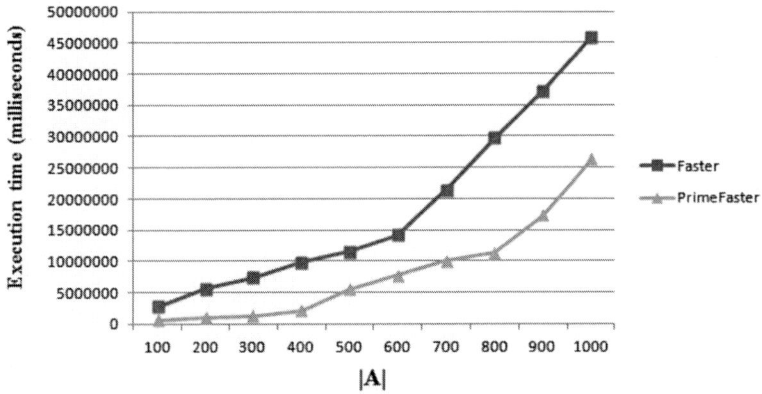

Test 5: In this test, we used formal contexts in the worst case. A formal context in the worst case is a triplet (O,O,R) with O a set of objects and R is a binary relation defined in O×O. The context size is varied between 10 and 50.

Figure 5: Diagram graph with contexts in the worst case

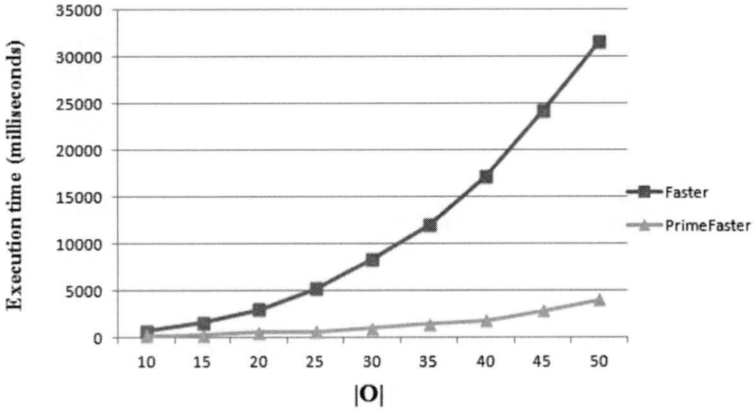

From Figure 5, we see that the run-time of the PrimeFaster algorithm is lower than that of the Faster algorithm. We also note that the gap between the two curves increases with increasing density of the formal context. We find that our improved version of the Faster algorithm (PrimeFaster) can reduce the execution time required for the extraction of formal concepts when the formal context becomes dense.

Test 6: In this test, we evaluated our algorithm using a real database Abscisic Acid Signaling Network Data Set (Jerry and Jenkins 2008). We performed 10 tests for the same number of objects (21) and the same number of attributes (43) each time by varying the density of the formal context.

Figure 6: Diagram graph with contexts in the worst case

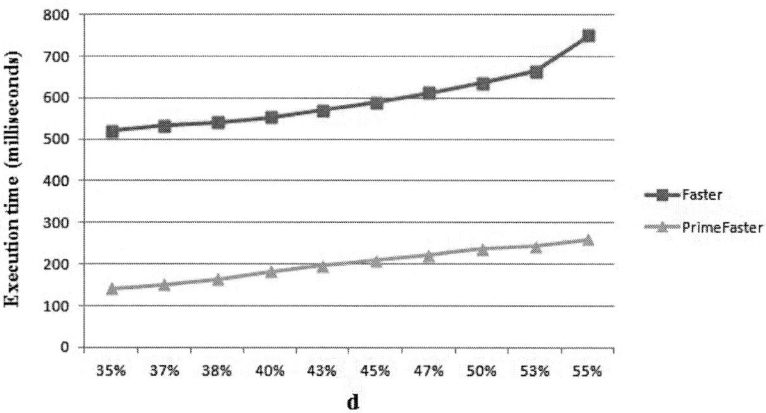

With experiments on the database Abscisic Acid Signaling Network Data Set (Jerry and Jenkins 2008), we have demonstrated the contribution of our work compared to that of the Faster algorithm (Choi 2006). Thus, we observe as is shown in Figure 6 that the run-time of our algorithm is always lower than that of the Faster algorithm. We also find that the gap between the two curves increases with increasing density of the formal context. This is explained by the number of concepts generated from the binary relation and the number of computing intersections. We conclude that our algorithm PrimeFaster performs well with dense formal contexts.

Conclusion

In this paper, we presented some algorithms of formal concepts extraction and some works that used prime numbers. Then, we presented our contribution which consists of defining a new version of the Faster algorithm (Choi 2006).

We have shown that the use of prime numbers in the encoding of objects and attributes of the formal context and the use of GCD and LCM in the comparison between intentions and extensions, allowed us to extract all formal concepts from a binary relation with a lower cost.

We have shown, experimentally, that PrimeFaster algorithm performance increases with an increase in data. The limitation of our work is related to the use of prime numbers for encoding objects and attributes when these prime numbers are very big.

References

Bjoem, K. (2006) "FooCA-Web Information Retrieval with Formal Concept Analysis", Dipl-Inform, Verlag Allgemeine Wissenschaft, Muhltal.

Bordat, J. P. (1986) "Calcul pratique du treillis de Galois d'une correspondance", Mathématiques, Informatique et Sciences Humaines, tome 96, pp. 31-47.

Carpineto, C. and Romano, G. (1996) "Information Retrieval through Hybrid Navigation of Lattice Representations", International Journal of Human-Computer Studies, vol.45, n. 5, pp. 553-578.

Chang, Y. I. Yang, B. I. and Yeh, W. H. (2001) "A Bit-Pattern-based Matrix Strategy for Efficient Iconic Indexing of Symbolic Pictures", The National Sciences Council of Republic of China under Grant No. NSC-87-2213-E-110-014.

Chein, M. (1969) "Algorithme de recherche des sous-matrices premières d'une matrice", Bull.Math.Soc.Sci. Roumanie, pp. 21-25.

Cheng, S. H., and Radha, G. S. V (2006) "Query Optimization Techniques for XML Database", International Journal of Information and Communication Engineering, pp. 381-388.

Choi, V. (2006) "Faster Algorithms for Constructing a Concept (Galois) Lattice", CLA.

Farhat, M., Harrathi, F. and Gammoudi, M. M. (2012) "A based on Formal Concept Analysis method for semi-structured query reformulation", The 5th International Conference SIIE, Djerba (Tunisia), pp. 189-193.

Gammoudi, M. M. (2005) "Décomposition conceptuelle des relations binaires et ses applications", Habilitation Universitaire en Informatique. Université de Tunis el Manar- Faculté des Sciences de Tunis.

Gammoudi, M. M. and Nafkha, I. (2002) "A formal method for inheritance graph hierarchy construction", Information Sciences, Elsevier Science,Vol. 140, No. 3, pp. 295-317.

Ganter, B. (1984) "Two basic algorithms", Preprint 831, Technishe, Hoshscule, Darmstad, 28.

Ganter, B. and Wille. R (1999) "Formal Concept Analysis", Mathematical Foundations, Springer-Verlag, ISBN: 3-540-62771-5, 284.

Godin, R., Rokia, M. and Hassan, A. (1995) "Incremental Concept Formation Algorithms Based on Galois (Concept Lattices)", Computational Intelligence, vol. 11, n2, pp. 246-267.

Jerry, W. J. and Jenkins, Ph. D. (2008) "Abscisic Acid Signaling Network Data Set", Systems Biology and Bioinformations Group, CFD Research Corporation, 215 Wynn Drive, Huntsville, AL 35805.

Karam, G., Mosab, H. and Mohammed, J. Z (2008) "An effective approach for frequent sequence mining via prime-block encoding", Journal of computer and System Sciences, Volume 76, Issue 1, pp. 88-102.

Kuznetsov, S. O. (2004) "Machine Learning and Formal Concept Analysis", In P. Eklund (Ed.), Concept Lattices: Second International Conference on Formal Concept Analysis, LNAI 296, pp. 287-312.

—. (1993) "A fast algorithm for computing all intersections of objects in a finite semi lattice", Automatic Documentation and Matjematical Linguistics, 27(5), pp. 11-21.

Kuznetsov, S. O. and Obiedkov, S. (2002) "Comparing performance of algorithms for generating concept lattices", Journal for Experimental and Theoretical Artificial Intelligence (JETAI).

Krajca, P., Outrata, J. and Vychodil, V. (2010) "Advances in algorithms based on CbO", Concept Lattices and Their Applications (CLA), Sevilla.

Mohammad, N. S., Norwati, M., Nasir, B. S. and Ali, B. M. (2008) "Maximal Frequent Itemsets Mining Using Database Encoding", The World Congress on Engineering Vol I WCE, London, U.K.

Norris, E. M. (1978) "An algorithm for computing the maximal rectangles in a binary relation", Revue Roumaine Mathématiques Pures et Applications, vol. XXIII.

Nourine, L. and Olivier, R. (1999) "A Fast Algorithm for Building Lattices", Information Processing Letters, vol. 71, pp. 199-204.

Priss, U. (2003) "Formal Concept Analysis in Information Science", 1-22.

Simon, A. (2009) "In-Close, a fast algorithm for computing formal concepts", 17-th International Conference on Conceptual Structures (ICCS), Moscow.

Vilem, V. (2008) "A New algorithm for computing formal concepts", In: Trappl, R. (ed.), Cybernetics and Systems, Proc. 19th EMSCSR.

CHAPTER SIX

SEGMENTATION OF IMAGES BASED ON THE STATISTICS OF RANKS OF THE LEVELS OF GREY

ZOHRA AMEUR, MINA BOULIFA, RAMY NEKKACHE AND SOLTANE AMEUR

In this work, we present a method of segmentation of images based on the statistics of ranks of the levels of grey. The adopted methodology consists in attributing to every pixel of the image a code according to the level of grey and the type of scanning of its neighborhood, then in estimating the frequency of appearance of the codes so obtained. The latter will train by definition the ranks of levels of grey and highlight the local structure of the distribution of the levels of grey in the image. Through this study, we highlighted the efficiency of the characterizing regions by means of statistics of ranks of levels of grey both for the regions and for the discrimination of textures. We took into account the relation that binds the spatial positions and the ranks of the levels of grey of pixels in a given neighborhood. Our method allows us to rerelease the local structure of the distribution of levels of grey in the image. The texture is then described by calculating the frequencies of appearance of these basic structures in a region.

This new description of texture allows the implementation of the discrimination and segmentation of textures. This method was applied to second generation Meteosat images. For that purpose, the raw images were transformed into images coded in an intermediate space so that the textural information is revealed. The segmentation creates the various classes of ground, sea and clouds with a good rate of classification.

Introduction

To analyze an image, we generally proceed in three stages: the segmentation, the description and the interpretation. The segmentation is the basic stage; it involves dividing the original image into homogeneous regions or in detecting the borders of these regions. These results are then supplied in the last two stages to describe objects and interpret the information contained in the image. Regions can be characterized in several ways: level of grey, colors, textures, and geometry. The purpose is then to determine regions corresponding to objects or to parts of objects of the image. In practice, the forms of objects are very varied and the perfect segmentation of an image is often difficult to obtain. This work involves segmenting images by using a technique based on the statistics of ranks of the levels of grey. The purpose is then to find the ranks of configurations of the levels of grey in a window to characterize regions or textures.

Generally, a satellite image allows for observation of the morphological properties of cloudy masses in the diverse scales defined in meteorology. These properties are the shape, the consistency and the texture. Furthermore, the multispectral information that the satellite provides allows analysis of the optical phenomena and the radiative exchanges occurring in the system Earth – atmosphere.

Discussion in the literature of the treatment of images has shown that the texture played a crucial role for the resolution of these questions (Unser 1986). As we are particularly interested in the texture of images, we adopted a methodology which involves choice. First, the approach suited the segmentation of this type of image. The following stages are a relevant choice of methods of analysis and classification for, on one hand, extracting the textural parameters and, on the other hand, establishing the criteria of discrimination and proceeding to the segmentation itself of the considered images.

The segmentation makes reference to the notions of difference and similarity perceived by the human visual system. These notions gave birth to two main approaches: the approach "Border" and the approach "Region". Let us notice that there are also the other less used approaches such as "the active outlines", "the templates", "the pyramidal models" and "the mixed models".

The term "Texture" is used to characterize the visual aspect of the surface of an object. This notion is frequently used in analysis of image. Indeed, to distinguish an image comprised of two different zones which have the same intensity of grey, it is useful to be interested in the motives characterizing these two zones. It is clear that the characterization of textures is another difficult problem, and it suggests that we always look for effective methods to analyze them and segment them. One of the main reasons is that there is no definition universally accepted which allows us to characterize the notion of texture. In this work, we present a method of segmentation of images based on the statistics of ranks of the levels of grey. This technique rests on a structural description of textures. The latters are considered as consisting of numerous basic motives which repeat periodically or randomly in a region. To highlight the local structure of the distribution of the levels of grey in the image, we took into account the relation which binds the spatial positions and the ranks of the levels of grey of pixels in a neighborhood.

The vectors of ranks have the effect of eliminating many configurations with regard to the direct use of the levels of grey. Nevertheless, the local information remains insufficient because of the number of pixels used for the coding. This method, initially elaborated for 4 – neighborhood, can be spread to 8 – neighborhood and 16 – neighborhood, this to take into account the anisotropy of the texture.

Developed Method

The ways which also allow for characterizing the texture are the approaches based on the evaluation of the vectors of ranks. These approaches involve coding every pixel of the image by crossing its neighborhood following appropriate paths. Although rather complex to formulate, the vectors of ranks are sensitive to the crossed directions. Consequently, they are capable of detecting the influence of the direction of the texture. This method is based on the statistics of ranks of the levels of grey. Translated in the form of an algorithm, it contains briefly the following stages:

The beginning

Step 0: initialization

- Acquisition of the size of the image;
- Choice of the size of Ω, window in which we calculate the frequencies of vectors of rank;
- Choice of the number of classes Nc (every class is represented by a color).

Step 1: Calculation of texture characteristics

- Determination of Vr row vectors (i, j) for each pixel for a given neighborhood;
- Coding vectors of ranks;

- Evaluation of the frequency of occurrence of vectors of rows in each window (Ω) centered at (i, j).

Step 2: Classification of pixels into Nc classes using a classification algorithm.

Step 3: Save the image Nc resulting in different colors.

End.

Neighborhood of a Pixel

A pixel I (i, j) has four horizontal and vertical neighbors that form what is called the 4-neighborhood of I (i, j). If we consider a pixel as an elementary square, the pixels having a common side with I (i, j) are called the 4-neighbors of (i, j).

Figure 2: The horizontal and vertical neighbors dimensions

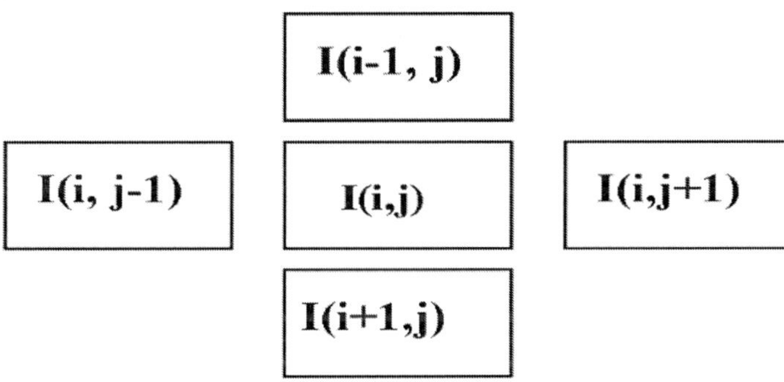

The Vector of the Levels of Grey

For every pixel I (i, j), the vector of the levels of grey of its 4 – nearby is defined by:

$$v(i,j) = \begin{bmatrix} f(i-1,j) \\ f(i,j-1) \\ f(i,j+1) \\ f(i+1,j) \end{bmatrix}$$ Where f(I,j) is grey level of the pixel I(i,j)

The writing of this vector can be simplified by using the indices of spatial positions. So, we shall have:

$$v(i,j) = \begin{bmatrix} f0 \\ f1 \\ f2 \\ f3 \end{bmatrix} = (f0, f1, f2, f3)^T$$; (T is the transposing vector)

Example: let a region of an image consists of 9 pixels

Figure 3: Calculation of the vector of the levels of grey

220	128	30
10	p(i, j)	128
39	120	240

$$v(i,j) = \begin{bmatrix} 128 \\ 10 \\ 128 \\ 120 \end{bmatrix}$$

The local information on the arrangement of the levels of grey in 4 – neighborhood is contained in the vector of the levels of grey. If we consider the combinations of the levels of grey of 4 – neighborhood as being a pattern, it gives us 256^4 = 4 294 967 296. To reduce this huge amount of information, we use the concept of rows of grey levels.

Vectors of the Ranks of the Levels of Grey

For every vector V (i, j) of level of grey, we can make correspond a vector of rank R (i, j) defined by:

$$R(i,j) = \begin{bmatrix} r0 \\ r1 \\ r2 \\ r3 \end{bmatrix} = (r0, r1, r2, r3)^T$$

Where r_i, i =0, 1, 2, 3, is the level rank of grey f_i in V (i, j), with $0 \leq r_i \leq 3$.

The vectors of the levels of grey can be described by following both manners:

- The vectors of different ranks.
- The vectors of general ranks.

The Vectors of Different Ranks

Let be fi, i= 0,.., 3; The level of grey of each of 4 vertical and horizontal nearby pixels (f0 corresponds to the level of grey of the pixel the position of which is 0). We list fi according to an increasing order. The row, i = 0… 3, gives the order of fi. So if we have f0 ≤ f1 ≤ f2 ≤ f3 , we shall have: r0 = 0; r1 = 1; r2 = 2; r3 = 3. We resume the previous example (Figure 3) to highlight better the principle of coding:

f0 =128, f1 = 10, f2 = 128, f3 = 120.
f1 ≤ f3 ≤ f0 ≤ f2
r1 = 0; r3 = 1; r0 = 2; r2 = 3.

Let us notice that the pixels of position 0 and 2 have the same level of grey, and different ranks: if fi = fj and i < j, then rj = ri + 1.

Unlike the vector of different ranks, the vector of general ranks allows us to distinguish the cases where the levels of grey are equal or different.

Let fi, i = 0,…,3, the level of grey of each of 4 vertical and horizontal nearby pixels (f0 corresponds to the level of grey of the pixel the position of which is 0). We list fi according to an increasing order. If fi = fj then ri = rj.

In the example of Figure 4, the position 0 has the smallest level of grey (f0 = 28); thus the rank of this position will have for value r0 = 0. Furthermore, the position 2 has the same level of grey as the position 0; thus r2 = r0 = 0. The position 1 will receive the rank 1 (r1 = 1); the position 3 will receive the rank 2 (r3 = 2).

Figure 4: Calculation of the vector of the levels of grey

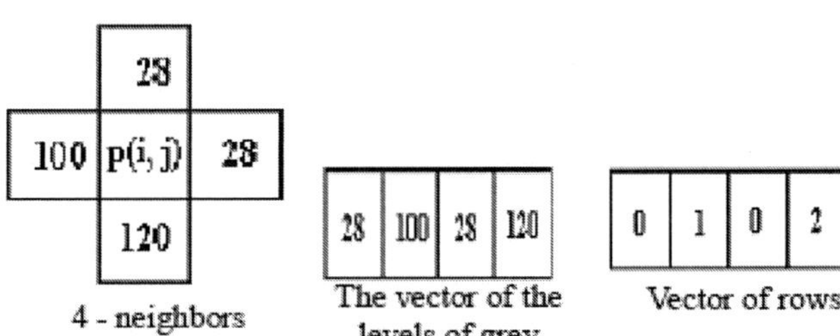

Coding of the Vector of Ranks

Every vector of rank corresponds to a pattern, and to discriminate between them, we attribute a code for every vector of ranks. This code is defined as follows:

$$Vr(i,j) = r_0 * 4^0 + r_1 * 4^1 + r_2 * 4^2 + r_3 * 4^3$$

The number of patterns depends on the size K of the neighborhood. The more K is important, the higher the number of motives perceived, but the latter is limited to the size of K. During the coding of the vectors of different ranks, we should have 24 possible combinations different from code.

Whereas during the coding of the general vectors, we should have 75 combinations different from possible codes. For K- neighborhood, the number of motives is:

- For a vector of the different ranks: Nvr = K * (K-1)…2*1=k!
- For a vector of the general ranks:

$$N_{v\ r} = \sum_{i=1}^{K} N_i(K) \qquad / \ N_1(K)=1$$

$$N_i(K) = i^K - \sum_{j=1}^{i-1} (\binom{i}{j}) * N_j(K) \qquad / \ (2 \le i \le K)$$

$$\binom{i}{j} = \frac{i!}{j!(i-j)\ !}$$

Remark: Although the vector of ranks eliminates many configurations with regard to the direct use of the levels of grey, the number of possibilities quickly becomes very important when the size of the neighborhood increases.

Table 1: Number of motives for K-NEIGHBORHOOD

K	1	2	3	4	5	6
Different rows	1	2	6	24	120	720
General rows	1	3	13	75	541	4683

Table 2: Coding of the vectors of general ranks (rank 75)

$Comb_i$	$V_r(i)$	$[r_0\ r_1\ r_2\ r_3]$	$Comb_i$	$V_r(i)$	$[r_0\ r_1\ r_2\ r_3]$	$Comb_i$	$V_r(i)$	$[r_0\ r_1\ r_2\ r_3]$
0	0	[0000]	25	57	[1230]	50	120	[0231]
1	1	[1000]	26	64	[0001]	51	129	[1002]
2	4	[0100]	27	65	[1001]	52	132	[0102]
3	5	[1100]	28	66	[2001]	53	133	[1102]
4	6	[2100]	29	68	[0101]	54	134	[2102]
5	9	[1200]	30	69	[1101]	55	135	[3102]
6	16	[0010]	31	70	[2101]	56	137	[1202]
7	17	[1010]	32	72	[0201]	57	141	[1302]
8	18	[2010]	33	73	[1201]	58	144	[0012]
9	20	[0110]	34	74	[2201]	59	145	[1012]
10	21	[1110]	35	75	[3201]	60	146	[2012]
11	22	[2110]	36	78	[2301]	61	147	[3012]
12	24	[0210]	37	80	[0011]	62	148	[0112]
13	25	[1210]	38	81	[1011]	63	152	[0212]
14	26	[2210]	39	82	[2011]	64	156	[0312]
15	27	[3210]	40	84	[0111]	65	161	[1022]
16	30	[2310]	41	88	[0211]	66	164	[0122]
17	33	[1020]	42	96	[0021]	67	177	[1032]
18	36	[0120]	43	97	[1021]	68	180	[0132]
19	37	[1120]	44	98	[2021]	69	198	[2103]
20	38	[2120]	45	99	[3021]	70	201	[1203]
21	39	[3120]	46	100	[0121]	71	210	[2013]
22	41	[1220]	47	104	[0221]	72	216	[0213]
23	45	[1320]	48	108	[0321]	73	225	[1023]
24	54	[2130]	49	114	[2031]	74	228	[0123]

Table 2 shows the various combinations of possible codes for vector of general ranks. In the table, $Comb_i$ is the number of the combination of the code, $V_r(i)$ is the value of the code, [r0 r1 r2 r3] the vector of ranks.

Capacity of Detection of the Orientation of Textures by Vectors of Ranks

The vectors of ranks supplies us the local patterns for every position of pixel. These patterns characterize the texture by generating the statistics of these patterns in a region or by constituting the rule of organization of these patterns in the region. The local direction of the change of levels of grey can be deduced from the vector of ranks. This is highlighted by Figure 5. The arrow in red indicates the direction according to which the level of grey increases in the case of two internal pixels to 4 – neighborhood.

Figure 5: Motive corresponding to four different directions

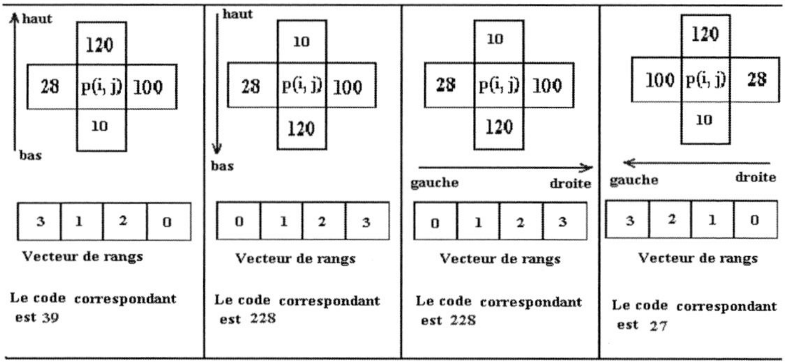

Calculation of the Frequencies of Appearance of the Vectors of Ranks in every Window (Ω) centered there (I, J)

For every pixel (i, j), (i = 1, Count lines 1; j = 1, Count columns 1), we choose a window centered on this one. We calculate on this window the frequencies of appearance of the vectors of ranks which we allocate to this pixel. Let us take an example to illustrate the principle.

Figure 6: Calculation of the frequencies of appearance of the vectors of ranks

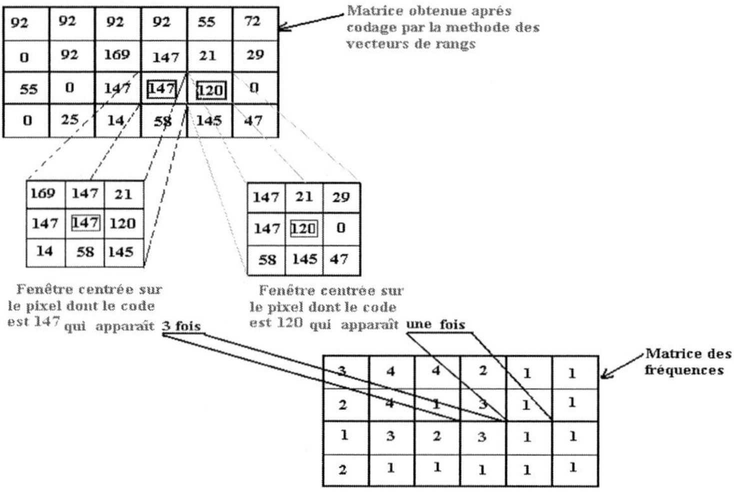

To show the capacity of characterization of a texture by the frequencies of the vectors of ranks, we took two textures from the album of Brodatz and we calculated for each of them the frequency of the vectors of ranks 75.

These textures are represented by Figure 7. The texture described by the Figure 7.a contains outlines in the horizontal direction and in the vertical direction.

The texture given by Figure 7.b rather consists of outlines in the diagonal direction.

Figure 7: Representation of two textured images

a. Tissue 1 b. Tissue 2

On the Figure 8, we find peaks of staff of vectors of ranks. Among these peaks, there are the codes 0 (corresponds to the vector of ranks [0000]). This pattern informs us that our image (Figure 7.a) consists of several homogeneous regions.

Figure 8: Frequencies of the vectors of ranks 75 of tissue 1

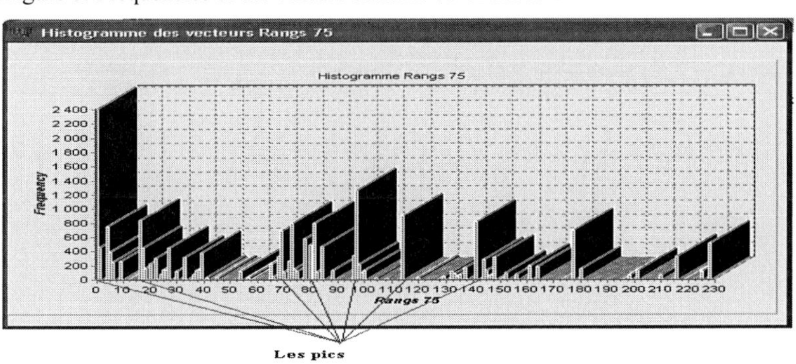

We notice that in the histogram represented by Figure 9, there are several peaks. It is due to the fact that the image represented by Figure 7.b, is trained by patterns having either abrupt changes of direction or trained by random patterns.

Figure 9: Frequencies of the vectors of ranks 75 of tissue 2

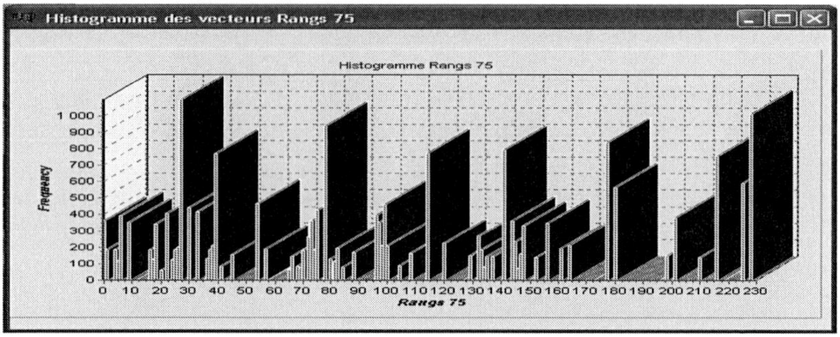

The vectors of ranks supply the local patterns for every position of pixel. With these local patterns, we can characterize the texture by calculating the statistical properties of these patterns in a region or by looking for the rule of organization of these patterns in the region.

We chose to characterize the texture by calculating the frequencies of the patterns in a region of the image. The table of the frequencies of these patterns serves then to characterize the texture.

Let us consider a region Ω formed of N pixels and let k=0, 1... Nm-1 with Nm=75 for the general ranks. The frequency F (k) of the pattern k, is defined as being:

$$F(K) = \frac{1}{N} \sum_{(i,j)\in\Omega} H(\nu_r(i,j)$$

$$H(\nu_r(i,j) \Rightarrow \begin{cases} 1 & s \ \ \nu_r^i(i,j) = c_k \\ 0 & s \ \ \nu_r^i(i,j) \neq c_k \end{cases} \quad c_k \in S$$

In this expression, the code of the vector at the rank k is with k=0, 1, Nm-1, and S is all the values of the possible codes, with:

$$\sum_{k=0}^{N_m-1} F(k) = 1$$

Figure 10: Calculation of frequency vector ranks

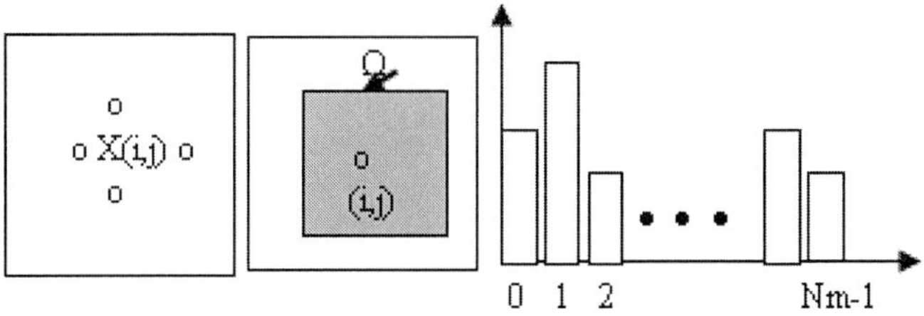

For every pixel of the image, a window centred on this one is chosen. We calculate on this window the frequencies of appearances of the vectors of ranks which we allocate to this pixel.

Tests and Results

Having realized our application, we established tests on satellite images taken on various channels (visible and infrared), and on images taken from the famous Google search engine.

Tests on the Images Meteosat

Our tests are made on three images illustrated by three figures: Figure 11, Figure 13, and Figure 15.

Segmentation parameters
Number of clusters: 6
Size of the window of analysis: 5 * 5
Algorithm of classification: K-Means
At the beginning of the processing: 16:22:18
At the end of the processing: 16:22:32

Figure 11: METEOSAT Image1 in infrared channel (18 juin 2006 9h00)

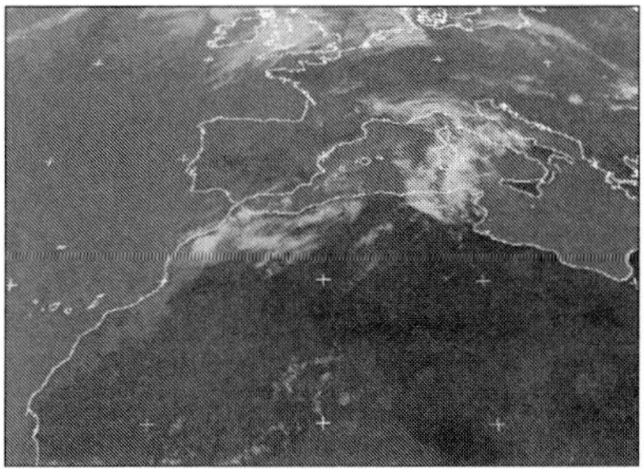

Figure 12: Segmented Image 1

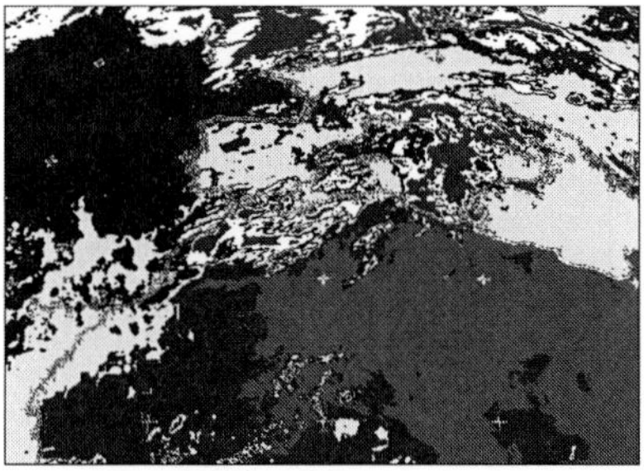

Segmentation parameters
Size of the window of analysis: 5*5
Number of clusters: 6
Algorithm of classification: K-Means
At the beginning of the processing: 16:32:31
At the end of the processing: 16:32:38

Figure 13: METEOSAT Image 2 in visible channel (18/06/2006 9h00). Copyright EUMETSAT

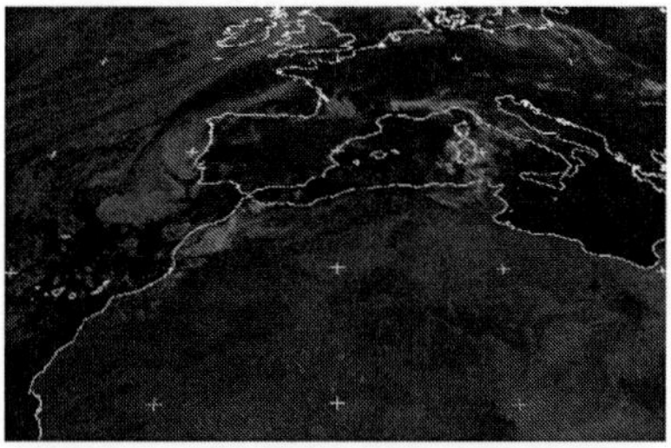

Figure 14: Segmented Image 2

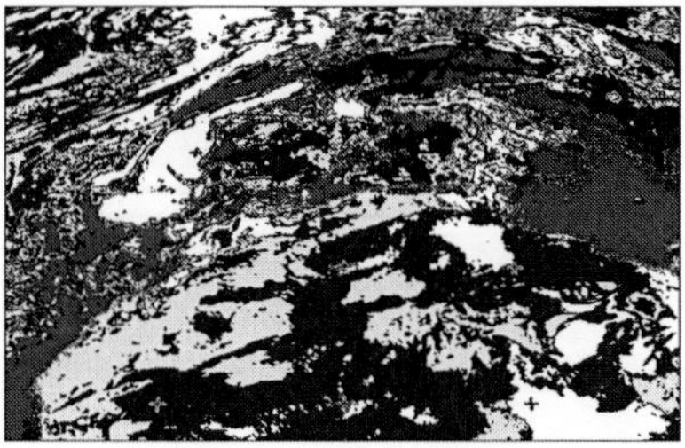

Segmentation parameters
Size of the window of analysis: 5 * 5
Number of clusters: 5
Algorithm of classification: K-Means
At the beginning of the processing: 16:37:32
At the end of the processing: 16:37:42

Figure 15: METEOSAT Image 3 in infrared channel

Figure 16: Segmented Image 3

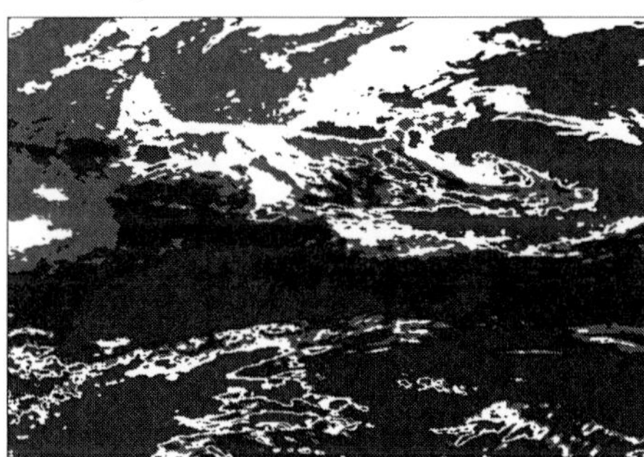

Tests on the Images of "Google Earth"

Segmentation parameters
Number of clusters: 7
Size of the window of analysis: 5 * 5
Algorithm of classification: K-Means
At the beginning of the processing: 17:06:58
At the end of the processing: 17:07:26

Figure 17: Image 4, Houari Boumediene international Airport

Figure 18: Segmented Image 4 (with 7 clusters)

Segmentation parameters (Figure 17)
Number of clusters: 6
Size of the window of analysis: 5 * 5
Algorithm of classification: K-Means
At the beginning of the processing: 17:13:27
At the end of the processing: 17:13:37

Figure 19: Segmented Image 4 (with 7 clusters)

Segmentation parameters
Number of clusters: 7
Size of the window of analysis: 5 * 5
Algorithm of classification: K-Means
At the beginning of the processing: 20:23:52
At the end of the processing: 20:24:03
Metrics of similarity: Chebyshev distance

Figure 20: Image5, aerial view of the port of Algiers (part)

Figure 21: Segmented Image 5

Segmentation parameters (Figure 20)
Number of clusters: 7
Size of the window of analysis: 5 * 5
Algorithm of classification: K-Means
At the beginning of the processing: 20:25:58
At the end of the processing: 20:26:22
Metrics of similarity: Euclidean distance

Figure 22: Segmented Image 5

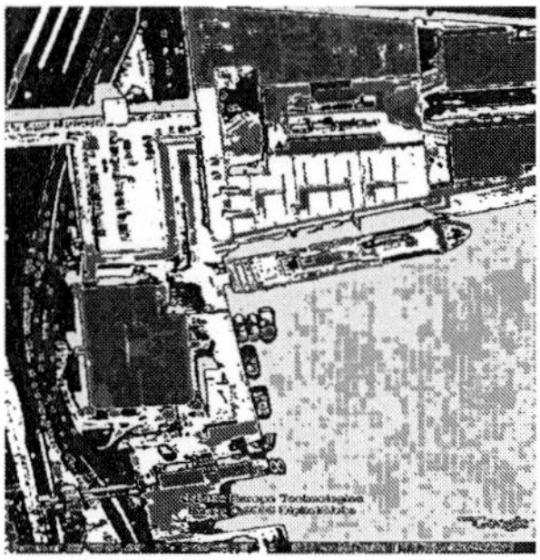

Segmentation parameters
Number of clusters: 7
Size of the window of analysis: 5 * 5
Algorithm of classification: K-Means
Metrics of similarity: Chebyshev distance

Figure 23: Image 6, aerial view of oil port of Arzow

Figure 24: Segmented Image 6

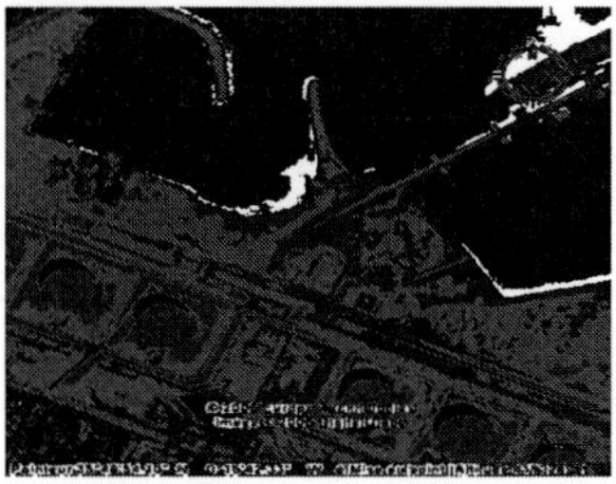

Discussion and Conclusion

The results obtained for both types of images – the images of the satellite of METEOSAT and the images of Google Earth – are very satisfactory. The result of the segmentation makes appear the various classes of ground, sea and clouds for the satellite images on one hand, and the various regions contained in the images of earth from Google on the other hand with a good rate of classification.

The choice of the parameters of segmentation is important for having a good segmentation. As an example, the result of the segmentation of image 5 (Figure 20) by using the metrics of similarity: Chebyshev distance, illustrated by Figure 21, is better than that obtained by using the metrics of similarity: Euclidian distance illustrated by Figure 22. Generally speaking, we noticed that the Euclidian distance discriminates well between our images.

We also noticed that the more we increased the number of classes, details seemed toengender the phenomenon of "on segmentation".

Our work consisted of implementing a new approach to the segmentation of images which is based on the statistics of ranks to be able to at best distribute the image in homogeneous regions. This algorithm takes into account the relation which binds the spatial positions and the ranks of the levels of grey of pixels in a given neighborhood, this with the aim of highlighting the local structure of the distribution of levels of grey in the image. The texture is then described by calculation of the frequencies of appearance of these basic structures in a region of the image.

The various realized tests showed the efficiency of this method applied to the images resulting from the satellite METEOSAT and to those supplied by the software "Google Earth".

The quality of the segmentation of the image depends on the used parameters, such as: the number of classes, the size of the window of analysis, the algorithm of classification and the metrics of similarity.

The segmentation of images, a recent and complex process, has not yet succeed in being equipped with a universal method capable of treating all the images. At the current stage of the evolution of these techniques, we are forced to proceed individually.

The segmentation represents at present a field of research characterized by fast and multiple evolutions with broadening perspectives including search engines for images, remote monitoring by computer, recognition of forms and other applications.

References

Ameur, Z. A new approach to segmentation of textured images, Magister thesis, UMMTO, 03 July 2001.

Ameur, Z., Ameur, S., Adane, A., Sauvageot, H., and Bara, K. Cloud classification using the textural features of Meteosat images, International Journal of Remote Sensing, 25(21), 2004, 4491-4503.

Ameur, Z. Image coding for a high-level segmentation: application to satellite images, Ph.D. thesis, UMMTO, September 25, 2005.

Ameur, Z., Adane, A., and Ameur, S. Texture image segmentation based on statistical rank directional, 3rd International Symposium on Image /Video Communications over fixed and mobile networks ISIVC (IEEE), 13-15 September 2006, Hammamet, Tunisia.

Ameur, Z., Adane, A., and Ameur, S. Classification images textures based on frequency vectors of rank, 4ème rd Conference International JTEA 2006, 12-14 May 2006, Tunis, Tunisia.

Bi, D., Asselin, J., de Beauville, P., and Slimane, M. Segmentation of textured images based on the vectors of ranks of gray levels and the Karhunen-Loeve, Four days of the French classification society, Vannes, France, 19 and 20 September, 1996.

Bara, K, Texture analysis in remote sensing: application to the segmentation of satellite images, Magister Thesis, UMMTO, 1998.

Cocquerez, J. P and Philipp, S. Image analysis: filtering and segmentation. Masson, Paris, 1995.

Conners, R. W., Conners, R. W., Trivedi, M. M., Harlow, C. A. Segmentation of a high resolution urban scene using texture operatures, Computer Vision, Graphics and Image Processing, 25(3), 1984, 273-310.

Desbois, M., Seze, G., and Szejwach, G. Automatic classification on METEOSAT imagery: Application to High Level Clouds, Journal of Applied Meteorology and Climatology, 21(3), 1982, 401-413.

Desbois, M. and Seze, G. Use of space and time sampling to produce representative satellite cloud classifications. Ann. Geophysic., 2(5), 1984, 599- 606.

Gueury, M. and Wijns, H. Preprocessing and data classification application to texture discrimination, Catholic University of Leuven, Faculty of Applied Sciences, France, 1994.

Kervrann, C. and Heitz, F. Unsupervised Segmentation of textured images: a statistical approach", IRISA, internal publication, No. 716, 1993.

CHAPTER SEVEN

PREDICTION OF THE FORM FUNCTION BY FUZZY SYSTEM TECHNIQUE

YOUSSEF NAHRAOUI, EL HOUCEIN AASSIF AND GERARD MAZ

Introduction

Several studies of characterization cover the acoustic scattering by targets of simple geometrical shapes (plate, tube, etc). These studies show that the acoustic resonances of a target, such as a tube, are related to its physical and geometrical properties (Hicking R 1964, Flax et al 1978, Murphy et al. 1978, Maze et al. 1983, Maze et al. 1985, Maze et al. 1991, Léon et al 1992, Ripoche et al. 1992, Doculto et al 1993, Veksler et al 1996, Morse et al 1998, Haumesser et al. 2002). Conversely, from these resonances, we can characterize the material of a target with a known geometry.

This work focuses on the spectrum of the backscattered acoustic pressure (or shape function FF) by an air-filled tube immersed in water. If the tube is excited by an acoustic plane perpendicular to its axis, circumferential waves are generated in the shell and water / shell interface (Talmant et al. 1988, Frisk et al. 1975). For certain frequencies, these circumferential waves form stationary waves on the circumference of the tube constituting resonance (Frisk et al. 1975). These resonances are observed in the spectrum of the backscattered acoustic pressure (Murphy et al. 1978, Maze et al. 1985). The circumferential waves belong to two types of waves that are equivalent to Lamb waves on a plate if the shell wall is thin: the symmetric (S_i) and antisymmetric (A_i) circumferential waves (i = 0, 1, etc: wave index) (Maze et al. 1985, Doculto et al. 1993, Talmant et al 1988). For a tube made of a given material, the reduced resonance frequencies of these waves essentially depend on the radius ratio b/a (Maze et al. 1991, Ripoche et al. 1992). The resonance modes *n*,

(The mode *n* is the number of wavelengths around the circumference), of a surface wave line up trajectories (n as function of the reduced frequency called Regge trajectories) (Maze et al. 1985, Izbicki et al. 1986).

According to Latif et al., for $i \geq 1$, these trajectories have reduced the cut-off frequencies, which depend on the physical characteristics of the tube (1999, 2000).

The aim of this paper is to compare the form function predicted by ANFIS techniques with those obtained analytically. The Wigner-Ville time-frequency distribution was used as a comparison tool to check the validity of the ANFIS network model to predict the form function (Latif, et al., 1999, 2000). The analysis of this time-frequency distribution takes into account both the time and frequency parameters, leading to the synthetic images that allow us to follow the evolution of the frequential content of a wave echo as a time function (Yen N et al. 1990, Claasen et al. 1980). Major benefits using ANFIS network include the excellent management of uncertainties, noisy data, and non-linear relationships. In this study, the ANFIS model predicts the form functions of tubes without the use of the analytical method for various radius ratio b/a. For each b/a, the input training data for ANFIS is a four-dimensional vector of the following form. W(t) = [FF(t-4) FF(t-3) FF(t-2) FF(t-1)].

The output training data corresponds to the trajectory prediction. s(t) = F(t). This model is able to predict the FF for copper tube of various radius ratio b/a between 0.9 and 0.99.

Backscattered Acoustic Signal from a Tube

Figure 1 shows the direction of a plane wave incident which will be scattered by an infinite cylindrical shell with a radius ratio b/a in a fluid medium. The axis of the cylindrical shell is identical to the z-axis of the cylindrical coordinate system(r, θ, z). The scattering of this plane wave is investigated through the solution of the wave equation and the associated boundary conditions.

The acoustic plane wave with frequency ω insonifies the cylindrical shell normally to the z-axis. The fluid 2, which fills the cavity of the shell, has a density ρ_2, and the velocity of longitudinal wave in this fluid is c_2. In general, the outer fluid, labelled fluid 1, is different, and its density is ρ_1. The velocity of wave is c_1.

In this study, an air-filled tube immersed in water is excited by an acoustic plane wave perpendicular to its axis. See Figure 1. The complex pressure P_{scat}^{∞} backscattered by the tube in a far field is the sum of the normal modes, which take into account the effects of the incident wave, the reflective wave {1}, surface waves in the shell {2} (whispering gallery waves, Rayleigh wave: equivalent to Lamb waves if the tube wall is thin) and an antisymmetrical interface Scholte A wave labelled also A_0 wave{3} connected to the geometry of the object Figure 2 (Maze et al. 1999. Waves {2} and {3} are the circumferential waves. In our case, the flexural waves A, A1 and A2 and the compression waves S_0, S_1 and S_2 are observed.

The general analytical form of the backscattered pressure P_{scat}^{∞} in a far field at normal incidence can be expressed as:

$$P_{scat}^{\infty}(\omega) = P_0 \frac{1-i}{\sqrt{\pi k_1 r}} \exp i(k_1 r - \omega t) \sum_{n=0}^{\infty} \varepsilon_n \frac{D_n^1(\omega)}{D_n(\omega)} \cos(n\theta). \qquad (1)$$

Figure 1: The geometry used to calculate the form function for an elastic tube

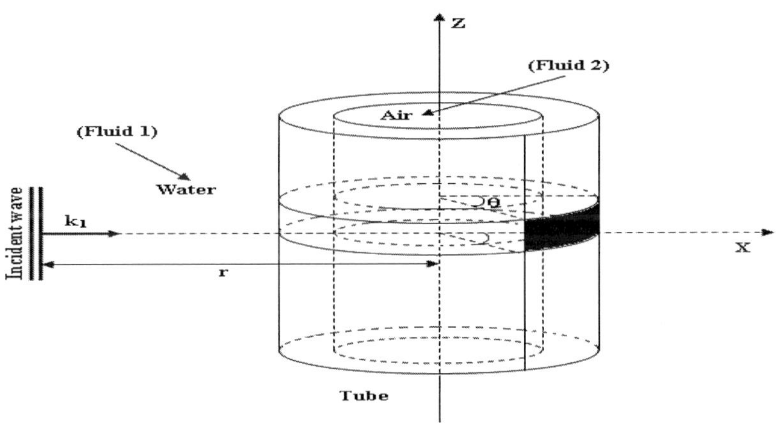

Figure 2: Mechanisms of the echo formation ({1}: specular reflection, {2}: circumferential shell)

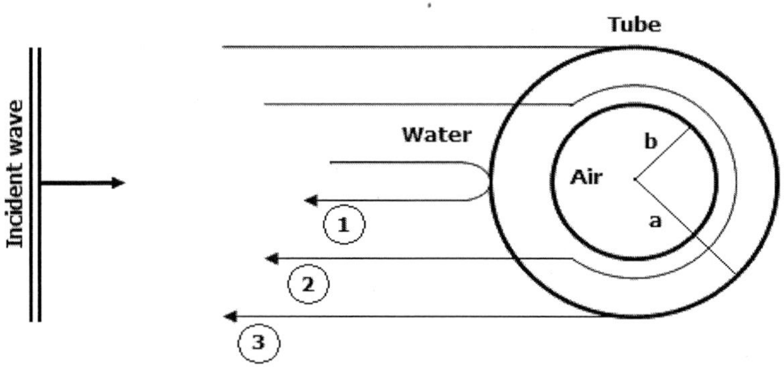

Where ω is the angular frequency, the following statements apply.

- k_1 is the wave-number with respect to the wave velocity in the external fluid.
- P_0 is the amplitude of the incident plane wave, r is the distance of pressure measurement.
- Θ is the azimuthal angle (Θ = 180° for the backscattering spectrum).
- $D_n^1(\omega)$ and $D_n(\omega)$ are determinants computed from the boundary conditions of the problem (continuity of stress and displacement of both interfaces). The terms of these determinants are given in ref 7 (in the present study, the normal incidence is only considered) and ε_n is the Neumann coefficient (ε_n =1 if n=0 and ε_n =2 if n≠0).

The physical parameters used in the calculation of the backscattered complex pressure are illustrated in Table 1.

Table 1: Physical parameters

	Density ρ(kg/m³)	**Longitudinal velocity C_L (m/s)**	**Transversal velocity C_T (m/s)**
copper	8930	4760	2325
Water	1000	1470	–
Air	1.29	334	–

Usually this backscattered pressure is presented as the form function FF defined by the relation (2) (Flax et al., 1977):

$$\left|\frac{P_{scat}^{\infty}}{P_0}\right| = \sqrt{\frac{a}{2r}}\, FF \tag{2}$$

This form function is the function of the reduced frequency $x_1 = k_1 a$ given by:

$$k_1 a = \frac{2\pi \nu a}{c_1} \tag{3}$$

where $\nu = \dfrac{\omega}{2\pi}$ is the frequency of a wave in Hz.

Figure 3 shows an example of the results computed for a copper tube (b/a = 0.94). The backscattered pressure spectrum is presented on Figure 3a. It is obtained from the computation of the form function FF (2). From the backscattered complex pressure, the impulse time response is calculated with an inverse Fourier transformation (4). The function h(ω) is the bandpass of transducers used in the experiments.

$$P_{scat}^{\infty}(t) = \frac{1}{2\pi}\int_{-\infty}^{+\infty} h(\omega) P_{scat}^{\infty}(\omega) e^{-i\omega t}\, d\omega \tag{4}$$

Examples of results are given in refs. 6 and 22. To obtain a resonance spectrum using "Resonance Scattering Theory" (RST), the theoretical studies indicate that it is possible to suppress the rigid background for the thick cylindrical shells, the soft background for the very thin shell and intermediate background for the other tubes (Flax et al. 1978, Murphy et al. 1978, Veksler et al. 1993). It is possible to obtain the resonance spectrum using another method developed in first time in experiment and applied to theoretical results: a new Fourier transformation is applied to the impulse time signal in which the specular echo (reflected echo) is suppressed and replaced by zeroes. This technique suppresses the real background, and the obtained spectrum is the backscattered resonance spectrum presented on Figure 3b (Pareige et al. 1989). The amplitude transitions (Figure: 3a) are related to the peaks observed (arrows) on Figure 3b (Flax et al. 1978, Murphy et al. 1978,Veksler et al. 1993).

These peaks are due to the resonances of modes *n*, (when or and?) they are connected with the propagation of circumferential waves (Murphy et al. 1978, Talmant et al. 1988). When a circumferential wave is generated on the shell, a resonance is established when the number of wavelengths on the circumference of the tube is an integer. This integer is the mode n. To plot Figure 3c, the resonance spectra are calculated for each value of *n* with the previous technique and associated in a plane "n-k1a" with an amplitude in grey level. The trajectories of the modes *n* of resonances in the same frequency domain (0<x1<200) are Regge trajectories (Maze et al. 1985, Izbicki et al. 1986). Each trajectory is related to a wave type (A, S_0, A_1). The trajectory in relation to the A1 wave has a cut-off frequency for the small values of the mode n indicated by an arrow on Figure 3c.

Wigner-Ville Time Frequency Distribution

The Wigner-Ville Distribution (WVD) plays a particularly important role in practice, such as energy conservation, time, frequency shift invariance, and preservation of the time duration and bandwidth (Claasen et al. 1980).

Time-frequency (TF) signal analysis and processing is concerned with the study of nonstationary signals or signals with time-varying frequency parameter leading to an image. This allows us to follow the evolution of the frequency content of acoustic echoes as a function of time (Hughes et al. 1993).

The WignerVille distribution (WVD) is a fundamental time-frequency distribution uniquely satisfies many desirable acoustical applications (Latif, et al., 1999, 2000).

The WVD of time complex signal u(t) displays the energy distribution of this signal in the time-frequency plane (5).

DWV is defined by (5):

$$W_u(t,\nu)=\int_{-\infty}^{+\infty} u\left(t+\frac{\tau}{2}\right)u^*\left(t-\frac{\tau}{2}\right)\exp(-i\,2\,\pi\,\nu\,\tau)\,d\tau \quad (5)$$

It represents the energy distribution of this signal in the time-frequency plane.

Where $\nu = \dfrac{\omega}{2\pi}$ is the frequency and *t* the time of the signal, $u^*(t)$ represents the complex conjugate of the signal $u(t)$.

Figure 3: (a) Form function spectrum, (b) Resonance spectrum and (c) Trajectories of modes n for an air-filled copper cylindrical shell (b/a=0.94) immersed in water. Arrows indicate the relation between transition (a) and peak (b). The reduced cut-off frequency of the A1 wave is indicated by a vertical line (c)

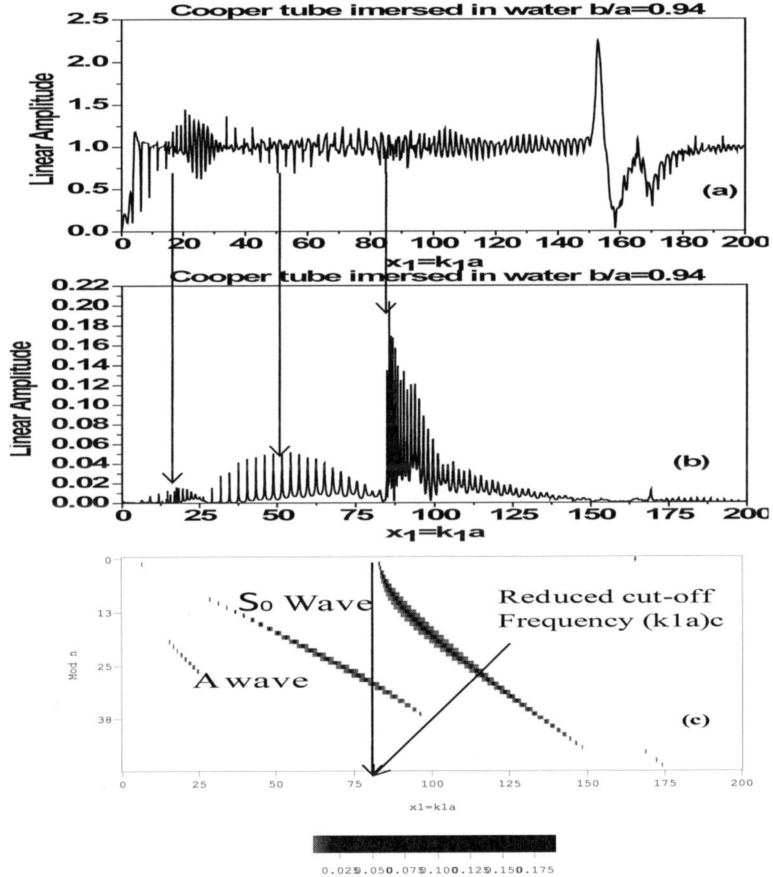

Neuro-Fuzzy Model

Fuzzy logic is a major research area that has been investigated by many scientists. It was proposed by Zadeh (1965) and has application in a large number of fields. Fuzzy logic, based on the fuzzy set theory, is an extension of Boolean since it takes into account intermediate values between "True or 1" and "False or 0", the only values recognized by the Boolean logic.

The Fuzzy Inference system (FIS) approach is based on the simple rules If-Then, which are easily manipulated to model the qualitative aspects of human knowledge and reasoning processes without having to recognise quantitative analysis. Basically, a fuzzy inference system is composed of five conceptual blocs, shown in Figure 4, as follows (Jang 1991):

(i) A rule base containing a number of fuzzy IF–THEN rules,
(ii) A database, which defines the membership functions of the fuzzy sets used in the fuzzy rules.
(iii) A decision-making unit, which performs the inference operations on the rules.
(iv) A fuzzification inference, which transforms the crisp inputs into degree of match with linguistic values.
(v) A defuzzification inference, which transforms the fuzzy results of the inference into a crisp output. These functional blocks are shown in Figure 4.

FIS implements a nonlinear mapping from its input space to the output space.

An adaptive network is a network structure whose overall input-output behavior is determined by a collection of modifiable parameters. The configuration of an adaptive network is composed of a set of nodes connected by direct links. Each node performs a static node function on its incoming signals to generate a single node output. The node function is a parameterized function with modifiable parameters.

It may be noted that links in an adaptive network only indicate the flow direction of signals between nodes and no weights are associated with these links. Readers are referred to Brown et al. (1994) for more details on adaptive networks. Jang (1993) introduced a novel architecture and learning procedure for the FIS that uses a neural network learning algorithm for constructing a set of fuzzy if-then rules with appropriate MFs from the stipulated input–output pairs. This procedure of developing a FIS using the framework of adaptive neural networks is called an adaptive neuro-fuzzy inference system (ANFIS).

ANFIS Architecture

The proposed architecture of the ANFIS is depicted in Figure 5b. Selection of the FIS is the major concern when designing an ANFIS to model a specific target system. Several types of FIS are reported in the literature (e.g. (Mamdani et al. 1975); (Flax et al. 1977); (Takagi et al. 1985) and are characterized by their consequent parameters only. The present study uses the Sugeno fuzzy model to generate fuzzy rules from a set of input-output data pairs (Takagi et al. 1985, Sugeno et al. 1988). This is because the conclusion of a fuzzy rule of this FIS is constituted by a weighted linear combination, and the parameters may be estimated by combination of the gradient descent method and the least squares estimate (LSE). To simplify the explanations, the fuzzy inference system under consideration is assumed to have two inputs (x and y) and one output *f*. For a first order of Sugeno fuzzy model, a typical rule set with base fuzzy if–then rules can be expressed as:

Rule 1: If x is A_1 and y is B_1, then $f_1=p_1 x+q_1 y+r_1$ (6)

Rule 2: If x is A_2 and y is B_2, then $f_2=p_2 x+q_2 y+r_2$ (7)

where A_1; A_2 and B_1; B_2 are the MFs for inputs x and y; respectively; p_1; q_1; r_1 and p_2; q_2; r_2 are the parameters of the output function. Figure 5a illustrates the fuzzy reasoning mechanism for this Sugeno model to derive an output function (*f*) from a given input vector [x, y].

The corresponding equivalent ANFIS architecture is presented in Figure 5b, where nodes of the same layer have similar functions. The functioning of the ANFIS is as follows:

- **Layer 1**: Each node in this layer generates membership grades of an input variable. The node output OP_1^i is defined by:

$$OP_1^i = \mu_{A_i}(x) \quad \text{for i=1,2}$$

or (8)

$$OP_1^i = \mu_{B_{i-2}}(x) \quad \text{for i=3,4}$$

Where x (or y) is the input to the node. A_i (or B_{i-2}) is a fuzzy set associated with this node, characterized by the shape of the MFs in this node and can be any appropriate functions that are continuous and piecewise differentiable such as generalized bell shaped functions.

Figure 4: Fuzzy Inference System with crisp output (Jang 1991)

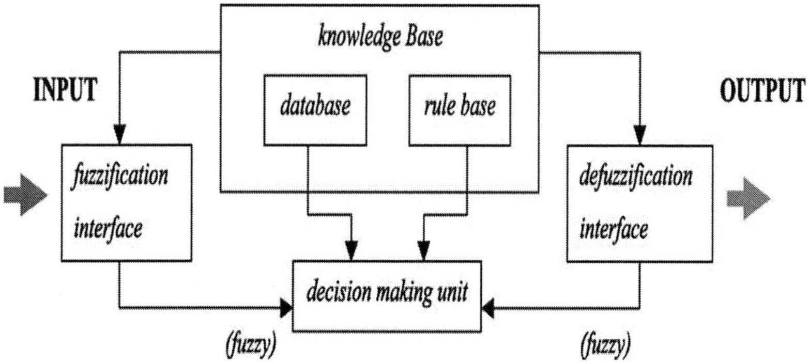

Assuming a generalized bell function as the MF, the output OP_1^i can be computed as:

$$OP_1^i = \mu_{A_i}(x) = \frac{1}{1 + (\frac{x - c_i}{a_i})^{2b_i}} \quad (9)$$

Where $\{a_i, b_i, c_i\}$ is the parameter set that changes the shapes of the MF with maximum equal to 1 and minimum equal to 0.

- **Layer 2**: Every node in this layer multiplies the incoming signals, denoted as π, and the output OP_i^2 that represents the firing strength of a rule is computed as:

$$OP_i^2 = \omega_i = \mu_{A_i}(x)\mu_{B_i}(y), i = 1, 2. \quad (10)$$

- **Layer 3**: The ith node of this layer, labeled as N; computes the normalized firing strengths as:

$$OP_i^3 = \varpi_i = \frac{\omega_i}{\omega_1 + \omega_2}, i = 1, 2. \quad (11)$$

- **Layer 4**: Node i in this layer computes the contribution of the ith rule towards the model output, with the following node function:

$$OP_i^4 = \varpi_i f_i = \varpi_i (p_i x + q_i x + r_i) \qquad (12)$$

Where ϖ is the output of layer 3 and $\{p_i, q_i, r_i\}$ is the parameter set.

- **Layer 5**: **The** single node in this layer computes the overall output of the ANFIS as:

$$OP^5{}_1 = Overall\ output = \sum_i \varpi_i f_i = \frac{\sum_i \omega_i f_i}{\sum_i \omega_i} \qquad (13)$$

Figure 5: (a) First-order Sugeno fuzzy model (b) ANFIS architecture(Jang 1993)

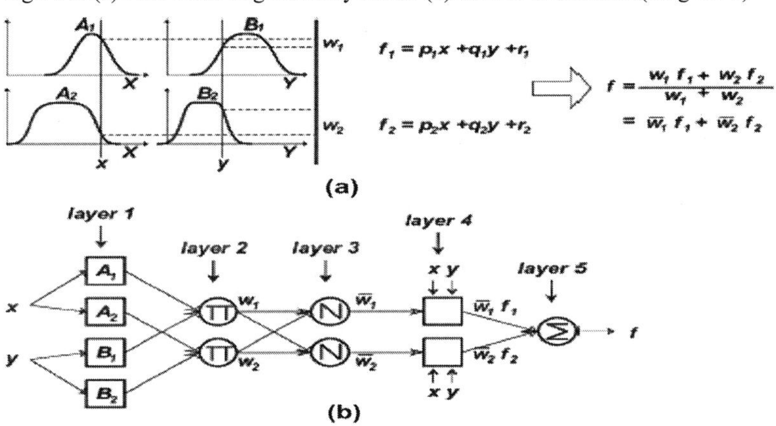

Estimation of Parameters

Two types of parameters characterize a fuzzy model: $\{a_i, b_i, c_i\}$ which determine the shape and distribution of the input fuzzy sets and $\{p_i, q_i, r_i\}$ which describe the overall output of the system (or linear models).

The basic learning rule of an adaptive network is based on a combination of the gradient descent method and the least squares estimate (LSE) to identify parameters.

From the proposed ANFIS architecture (see Figure 5), it is observed that given the values of the premise parameters, the overall output can be expressed as linear combinations of consequent parameters. More precisely, the output *f* can be rewritten as:

$$
\begin{aligned}
f &= \varpi_1 f_1 + \varpi_2 f_2 \\
&= (\varpi_1 x)p_1 + (\varpi_1 y)q + (\varpi_1)r_1 + (\varpi_2 x)p_2 + (\varpi_2 y)q_2 + (\varpi_2)r_2
\end{aligned}
\tag{14}
$$

This is linear in the consequent parameters (p1; q1; r1; p2; q2 and r2). As a result, the total number of parameters S in an ANFIS can be divided into two sets: S1= set of premise parameters and S2 = set of consequent parameters. Therefore, the hybrid learning algorithm can be applied directly, which combines the back propagation gradient descent and least squares method, and therefore, can be used for an effective search of the optimal parameters of the ANFIS. More specifically, in the forward pass of the hybrid learning algorithm, the node output goes forward until layer 4 and the consequent parameters are identified by the least squares method. In the backward pass, the error signal propagates backwards and the premise parameters are updated by gradient descent. As mentioned earlier, the identified consequent parameters are optimal, provided that the premise parameters are fixed.

Accordingly, the hybrid approach converges much faster since it reduces the dimension of the search space of the original back-propagation method. A detailed description of this algorithm can be found in Jang (1991).

Application of the Adaptive Neuro-Fuzzy Inference System (ANFIS) Networks to Predict FF

System modelling based on conventional mathematical tools (eg., differential equations) is not well suited for dealing with ill-defined and uncertain systems. By contrast, an ANFIS model employing fuzzy if-then rules can model the qualitative aspects of human knowledge and reasoning processes without employing precise quantitative analyses. The advantage of ANFIS networks is that they are capable of a linear and nonlinear modelling of systems.

In the present study, we use an ANFIS trained with the gradient descent (Jang, 91) back-propagation and last squares algorithm to achieve the learning ability to predict the form function for copper tubes (Jang 1991, Dariouchy et al. 2007).

ANFIS method constitutes a new approach of prediction of the form function for various radius ratio b/a of tubes included in the interval [0.4-

0.99]. With this method, we can predict the form functions for tubes of various radius ratio b/a and for various materials

ANFIS Model and Training Phase

ANFIS networks method training requires a set of form functions that are calculated by the analytical method or obtained by experiments.

In this work, the dataset constitute of the form functions is calculated by the analytical method. This dataset is divided into two sets. The first 1000 training data set was used for training the ANFIS, while the remaining 1000 checking data set was used for validating the identified model. The number of membership function is fixed to 2 MF, so the rule number is 16. The ANFIS used here contains a total of 104 fitting parameters, of which 24 are premise parameters and 80 are consequent parameters. The desired and predicted values for both training data and checking data are essentially the same in figures 7, 8 and 9. The input vector of the ANFIS is represented by the radius ratio b/a and the FF values for the preceding six indexes of dimensionless frequency:

$$(FF)_{x_1(i)} = ANFIS[\frac{b}{a},(FF)_{x_1(i-1)},(FF)_{x_1(i-2)},$$
$$(FF)_{x_1(i-3)},(FF)_{x_1(i-4)},] \quad (15)$$

Once the dataset is constructed, we train the ANFIS using the combination of last square error and the back-propagation gradient descent algorithm to predict the form function.

The training phase requires a few minutes. Once the network is trained, the parameters of membership function and consequence are fixed and will be used thereafter to predict any form function included in the *b/a* interval [0.9 – 0.99]. The last interval can be extended from 0.4 to 0.99 without loss of accuracy.

Selection of the optimal model

The performance of ANFIS models for training and testing data sets were evaluated according to statistical criteria, such as coefficient of correlation R, MAE, MRE, and SE. The selection of different models is done by comparing the errors of the ANFIS configuration, calculating the MAE, the MRE and the SE of the form function. The coefficient of correlation R and the determination R^2 of the linear regression are used as performance measures of the model between the predicted and the desired output. The different error measures and the coefficient of correlation are given by the following relations:

$$MRE=\frac{1}{N}.\sum_{i=1}^{N}\frac{|C_i-O_i|}{C_i}\times 100, \tag{16}$$

$$MAE=\frac{1}{N}\sum_{i=1}^{N}|C_i-O_i| \tag{17}$$

$$SE=\frac{\sqrt{\sum_{i=1}^{N}(C_i-O_i)^2}}{N-1}, \tag{18}$$

$$R=1-\frac{\sum_{i=1}^{N}(C_i-O_i)^2}{\sum_{i=1}^{N}(C_i-O_m)^2}, \tag{19}$$

Where N is the number of data, C_i and O_i are respectively the desired and the predicted values. O_m is the mean of the predicted values.

Results and Discussion

The ANFIS is trained by randomly using 50% of the dataset, whereas the remaining dataset 50% is reserved as a test base. The analysis is repeated several times. Indeed, the values of errors are measured for ANFIS network architecture, and the best configuration is found for a network with the number of membership function fixed to 2 MF, so the number of rules is 16. The MRE and the SE for the optimal model are respectively 1.9% and 6.10-4 k1a for b/a=0.97 Figure 6.

Figures 7 and 8 show the comparison between the form functions predicted by the ANFIS model on a test database and that calculated by the analytical method for the radius ratios b/a = 0.94 and 0.96. Figures 7 and 8 show a good agreement between the predicted and the calculated values of the form function.

However, the time-frequency representation of Wigner-Ville is used as a tool of comparison to check the validity of the ANFIS network model because the studied signals are not stationary. Using a time-frequency representation can allow us to follow the evolution of the frequency content of acoustic echoes as a function of time. Figure 9 shows the good agreement between the Wigner-Ville image of the predicted and the calculated form function for a copper tube of radius ratio b/a=0.94 and 0.96. Moreover, starting from the Wigner-Ville image of the predicted signal, the reduced cut-off frequency ([k1a]c) of the tube corresponding to the A_1 wave can be determined.

The circumferential asymmetric A_1 waves propagate around the circumference of the tube only for frequencies superior to the reduced cut-off frequency (Dariouchy et al. 2007). The reduced cut-off frequency values $([k_1a])_c$) are obtained from the Wigner-Ville image of the test database (b/a=0.94, 0.96) and analytically calculated are presented in Table 2.

Starting from the operating mode direction of this optimal model, we can make several things. In a direct operating mode of the model optimal, one can predict the form functions for tubes of various radius ratio b/a and for various materials.

In an inverse operating mode, the network model can go up to the radius ratio b/a, starting from an unknown form function for a given material and this by using the parameter of membership function of the optimal model. Generally, the ANFIS networks methods can contribute to the comprehension and the interpretation of the backscattered acoustic pressure by submerged tubes.

Figure 6: Correlation between calculated and predicted values of form function on a test dataset for radius ratio (a) b/a=0.94, (b) b/a=0.96

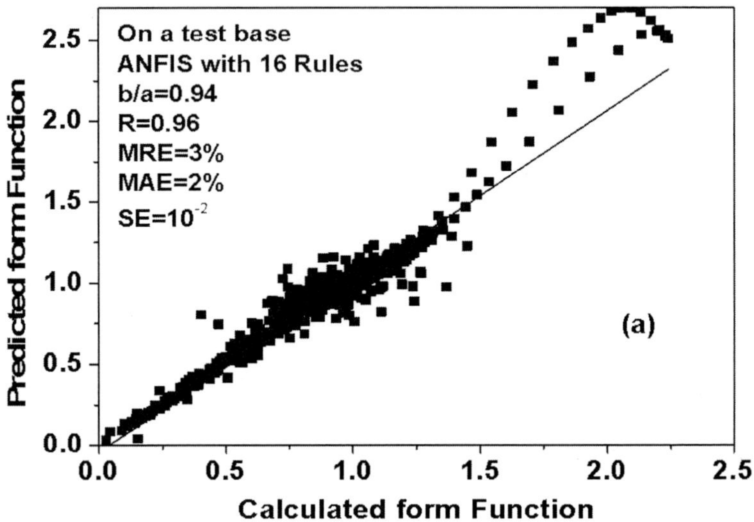

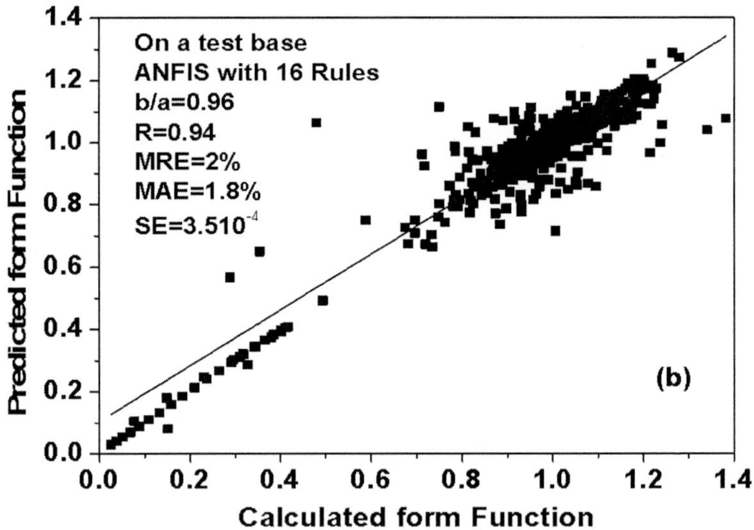

Figure 7: Form Function calculated by analytical methodand form Function predicted by ANFIS for b/a=0.94

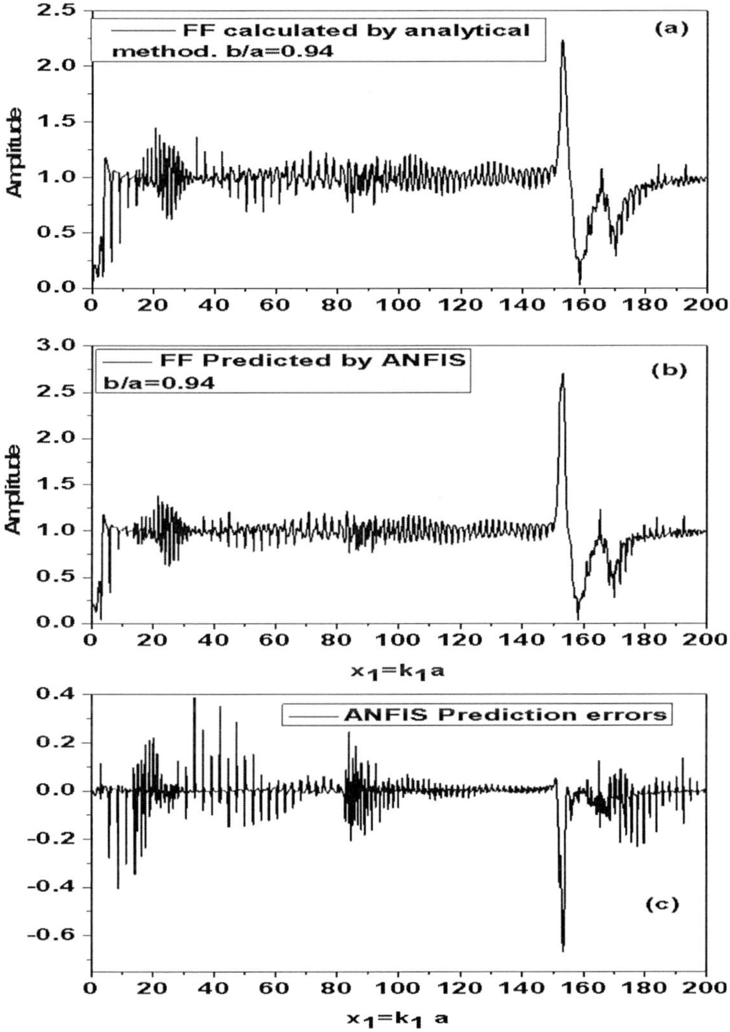

Figure 8: Form Function calculated by analytical method and form Function predicted by ANFIS for b/a=0.96

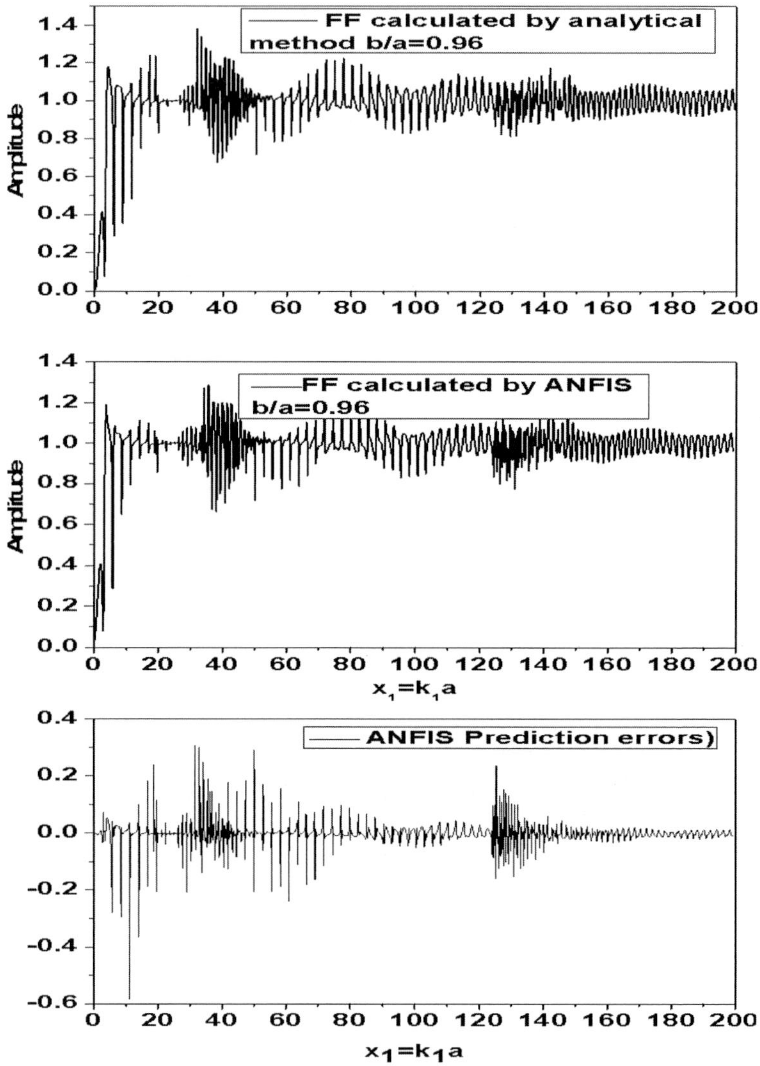

Figure 9: Comparison between the WV images for the form function calculated by the analytical method and predicted by the *ANFIS* method (b/a=0.94 and 0.96)

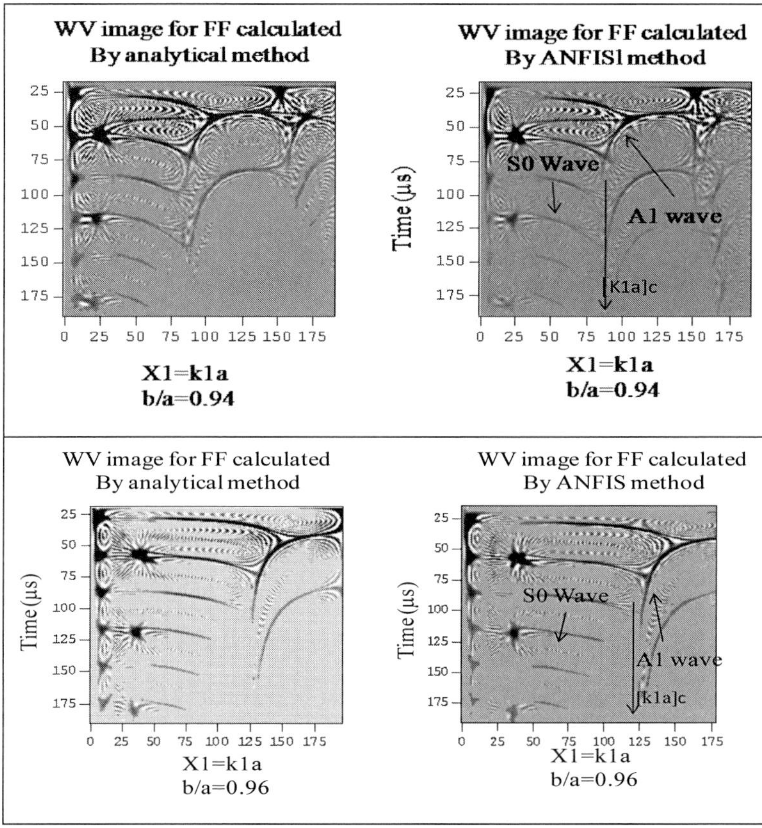

Table 2. Comparison between the reduced cut-off frequencies computed by eigen modes method and starting from the Wigner-Ville image

Circumferential wave A_1	$(k_1a)_c$ Estimated (WV) by analytical method	$(k_1a)_c$ Estimated (WV) by ANFIS
b/a=0.93	70.94±0.3	70.48±0.3
b/a=0.94	82.75±0.2	82.80±0.2
b/a=0.95	99.32±0.3	99.94.0±0.3
b/a=0.96	124.13±0.2	124.11±0.2
b/a=0.97	165.54±0.2	165.00±0.2

Conclusion

This paper presents the application of an *ANFIS* model to predict the form function for a copper tube. We can say that the results obtained with the new proposed approach are interesting and positive in prediction. This model is able to predict the FF with a Mean Relative Error (MRE) of about 2%. In this article, the ANFIS approach can be used as a new tool for non-destructive characterization of a thin elastic tube. To check the credibility of the ANFIS model in time and frequency domains, time-frequency representation of Wigner-Ville was used as a tool of comparison between the FF calculated by the analytical method and that predicted by the *ANFIS* technique.

The reduced cut-off frequencies estimated by the *ANFIS* method are in good concordance with those computed by the Eigen modes method. The comparison of the predicted and the calculated results indicates that the neuro-fuzzy method is suitable to predict the backscattered pressure. In this paper, we have worked on copper tubes but our *ANFIS* model can also be applied to various types of materials taking into account the material density as a relevant input of the model.

References

Brown, M., Harris, C., 1994. Neurofuzzy Adaptive Modeling and Control. Prentice Hall.

Claasen TACM, Mecklenbrauker WFG. The Wigner-Distribution tools for time-frequency signal analysis. Philips J. Res. Vol. 35, n° 3, 4, 5, 217-250, 276-300 and 372-389 (1980).

Décultot D., Lecroq F., Maze G., Ripoche J., Acoustic scattering from a cylindrical shell bounded by hemispherical endcaps. Resonance explanation with surface waves propagating in cylindrical and spherical shells, J. Acoust Soc. Am. 94, 2916-2923 (1993).

Dariouchy A, Aassif EH, Decultot D and Maze G. Acoustic Characterization and Prediction of the Cut-Off Dimensioless Frequency of an Elastic Tube by Neural Networks. IEEE Transaction on Ultrasonics, Ferroelectrics, and Frequency Control 54 (5), 1055-1064 (2007).

Frisk G.V., Dickey J.W., Überall H., Surface wave modes on elastic cylinders, J. Acoust Soc. Am. 58, 996-1008 (1975).

Flax L., Neubauer W. G., Acoustic reflection from layered elastic absorptive cylinders,J. Acoust. Soc. Am. 61, 307-312 (1977).

Flax L., Dragonette L.R., Überall H., Theory of elastic resonance excitation by sound scattering, J. Acoust Soc. Am. 63, 723-731 (1978).

Hickling R., Analysis of echoes from a hollow metallic sphere in water, J. Acoust Soc. Am. 36, 1124-1137 (1964).

Hughes DH, Marston PL. Local temporal variance of Wigner's distribution function as a spectroscopic observable: Lamb wave resonances of a spherical shell. J. Acoust. Soc. Amer. 94, 499-505 (1993).

Haumesser L., Décultot D., Léon F., Maze G., Acoustic scattering from a finite cylindrical shell at oblique incidence: Experimental identification along the shell length, J. Acoust Soc. Am. 111, 2034-2039 (2002).

Izbicki J.L., Maze G., Ripoche J., Influence of the free modes of vibration on the acoustic scattering of a circular cylindrical shell, J. Acoust Soc. Am. 80, 1215-1219 (1986).

Jang, J.-S.R., 1991. Rule extraction using generalized neural networks. In Proceedings of the fourth IFSA World Congress 4, 82–86.Volume for Artificial Intelligence.

—. 1993. ANFIS: adaptive network based fuzzy inference system. IEEE Transactions on Systems, Man and Cybernetics 23 (3), 665–683.

Léon F., Lecroq F., Décultot D., Maze G., Scattering of an obliquely incident acoustic wave by an infinite hollow cylindrical shell, J. Acoust Soc. Am. 91, 1388-1397 (1992).

Latif R., Aassif E.H., Maze G; Décultot D., Moudden Ali, Determination of group and phase velocities from time-frequency representation of Wigner-Ville, NDT & E International 32, 415-422 (1999) and Latif R., Aassif E.H., Moudden A., Décultot D., Faiz B., Maze G; Determination of the cutoff frequency of an acoustic circumferential wave using a time-frequency analysis, NDT & E International 33, 373-376 (2000).

Mamdani, E.H., Assilian, S., 1975. An experiment in linguistic synthesis with a fuzzy logic controller. International Journal of Man-Machine Studies 7 (1), 1–13.

Murphy J.D., Breitenbach E.D., Überall H., Resonance scattering of acoustic waves from cylindrical shells, J. Acoust Soc. Am. 64, 677-683 (1978).

Maze G., Acoustic scattering from submerged cylinders, MIIR Im/Re: Experimental and theoretical study, J. Acoust Soc. Am. 89, 2559-2566 (1991).

Maze G., Ripoche J., Visualization of acoustic scattering by elastic cylinders at low ka, J. Acoust Soc. Am. 73, 41-43 (1983).

Maze G., Izbicki J.-L., Ripoche J., Resonances of plates and cylinders: Guided waves, J. Acoust Soc. Am. 77, 1352-1357 (1985).

Maze G., Léon F., Ripoche J., and Überall H. Repulsion phenomena in the phase-velocity dispersion curves of circumferential waves on elastic cylindrical shells, J. Acoust. Soc. Amer. 105, 1695-1701 (1999) and Maze G., Touraine N., Baillard A., Décultot D., Latard V, Derbesse L., Pernod P., Merlen A., A0-wave and A-wave in cylindrical shell immersed in water: influence on the acoustic scattering, (1999) ASME, DECT, Las Vegas, Nevada, USA 12-16 sept 1999, Proceeding on CD-Rom.

Morse S.F., Marston P.L., Kaduchak G., High-frequency backscattering enhancements by thick finite cylindrical shells in water at oblique incidence: Experiments, interpretation and calculations, J. Acoust Soc. Am. 103, 785-794 (1998).

Pareige P., Rembert P, Izbicki J.-L., Maze G., Ripoche J., Méthode impulsionnelle numérisée (MIN) pour l'isolement et l'identification des résonances de tubes immergés, (digital impulse method for isolation and identification of resonances of immersed tubes), Physics Letters 135A, 143-146 (1989).

Ripoche J., Maze G., A new acoustic spectroscopy: The resonance scattering spectroscopy by the MIIR, "Acoustic Resonance Scattering", Redactor H. Überall, Editor Gordon and Breach, New York, 1992, (ISBN 2-88124-513-7), § 5, 69-103.

Sugeno, M., Kang, G.T., 1988. Structure identification of fuzzy model. Fuzzy Sets and Systems 28, 15–33.

Sun N. H., Marston P. L., Ray synthesis of leaky Lamb wave contributions to backscattering from thick cylindrical shells, J. Acoust. Soc. Amer. 91, 1398-1402 (1992).

Takagi, T., Sugeno, M., 1985. Fuzzy identification of systems and its application to modeling and control. IEEE Transactions on Systems, Man and Cybernetics 15 (1), 116–132.

Talmant M, Quentin G, Rousselot JL, Subrahmanyam JV, and Überall H. Acoustic resonances of thin cylindrical shells and the resonance scattering theory. J. Acoust. Soc. Am. 84, 681-688 (1988).

Tsukamoto, Y., 1979. An approach to fuzzy reasoning method. In: Gupta, M.M., Ragade, R.K., Yager, R.R. (Eds.), Advances in Fuzzy Set Theory and Application, North-Holland, Amsterdam, pp. 137–149.

Veksler N., Maze G., Ripoche J., Porochovskii V., Scattering of obliquely incident plane acoustic wave by circular cylindrical shell. Results of computations., Acta Acustica 82, 689-697 (1996) and Maze G., Léon F., Veksler N., Scattering of an obliquely incident plane acoustic wave by circular cylindrical shell. Experimental results, Acta Acustica 84, 1-11 (1997).

Veksler N., Resonance Acoustic Spectroscopy, Springer Series on Wave Phenomena, Springer-Verlag, (1993)Berlin, Heidelberg, New York, ISBN 3-540-55638-9.

Yen N, Dragonette L, K. Numrich S. Time-frequency analysis of acoustic scattering from elastic objects. J. Acoust. Soc. Am. 87, 2359-2370 (1990).

Zadeh L, "Fuzzy Sets," Inform. Contr., vol. 8,no. 3, pp. 338-353,1965.

CHAPTER EIGHT

COMBINATION OF FEATURE GENERATION METHODS FOR RECOGNITION OF NORMALIZED AND NOT NORMALIZED ISOLATED HANDWRITTEN DIGITS

ABDELJALIL GATTAL AND YOUCEF CHIBANI

In this chapter, the generation of features is an important recognition module in a particular and very serious aspect of the field of automatic recognition of normalized and not normalized isolated handwritten digits, which requires having a large database of digits. As the aim is to improve the performance of the recognition system, we are focused on combining different pertinent structural and statistical features generated from the digits. The steps of the proposed approach are as follows: First, we start by applying various structural feature generation methods which are generated from each handwritten digit image. Next, we use the uniform grid technique to increase the recognition rate. Then, the classification is done using SVM. Experimental results conducted on a large database of isolated handwritten digit NIST SD19 show that this combination is effective.

Introduction

In processing a sequence for automatic recognition of handwritten digits, one of the most important steps is to describe, in a discriminatory way, the digit to be processed. We can define the feature generation methods like the extraction problem of the most pertinent information starting from the image for a classification problem, i.e., that minimizes the intra-class variability and maximizes the inter-class variability (Devijver and Kittler 1982, Yan 2001, Kim et al. 2002). This pertinent information often takes the shape of a numeric value vector. The feature generation methods are many and varied. Each one has its own properties and can apply only to various contexts and meet to different conditions. Before choosing one method rather than another, it is important to know all the ins and outs of their use.

Many researchers have tried to improve the performance of recognition systems by studying various feature generation and classification methods. Selection of a feature generation method is the single most important factor in achieving high recognition performance. For this reason, the literature of handwritten digit recognition shows several kinds of features. For example, the first kind is based on the gray-level or binary values of all the pixels in an image, usually represented by number of pixels in the image.

The structural features are usually perceptual objects of the digit such as bends, end points, intersections, loops, measures of concavity, etc. (Pittner and Kamarthi 1999). Finally, the statistical features are results of global transformations of an image. Typical mathematical transformations include moments (Bailey and Srinath 1996), Fourier descriptors, and Wavelet/ Ridgelet transforms (Hu and Yan 1998).

In the following, we try to cite some algorithms according to mentioned kinds. The method of Cai and Liu (2001) presents an approach that integrates both statistical and structural information for the recognition of unconstrained handwritten numerals. This method was evaluated using 2,213 images from the CEDAR database and they reported a correct segmentation rate of 98.4%.

Heutte et al. (1998) propose a combination of seven different features to feed a linear discrimination-based classifier. This method was tested on 53,324 images of the NIST SD3 database and resulted in an accuracy rate of 94.8%.

In order to get more reliable systems, we have turned towards the combination of statistical and structural features.

We present in this chapter a generation process of features based on a combination of structural and statistical approaches (Bortolozzi et al. 2005, Yan 2001, Kim et al. 2002). This process mainly uses the global features, the Ridgelet transform, background and foreground features (Contour, skeleton) extracted from the images. The goal is then to obtain the best possible structural representation of the processed digits using these features. The goal of our approach is to bring together some feature generation methods whose objective is to find the best possible representative of the digit structure.

For this purpose, firstly, we draw up the uniform grid method to split the image. Then, we present a synthesis of some important properties of these structural methods. Experimental results conducted on a very large database are presented in order to find the best possible combination. Finally, a conclusion and discussion offuture works are presented in the last.

Region Sampling: Uniform Grid

In order to generate the most pertinent information and to increase the feature vector, we recall here the uniform grid method (Favata and Srikantan 1996). A uniform grid creates rectangular regions for sampling; each one has the same size and shape. The starting image is a binary image of size H x L. For each rectangular region, it is necessary to generate a feature vector. The positions of both horizontal and vertical grid lines are found using the following equation:

$$p_i = \left\lceil i \times \frac{k}{n} \right\rceil \qquad i = 1, 2, \ldots, n\text{-}1 \qquad (1)$$

Where the value p is the vector of line positions, n is the number of horizontal or vertical regions, and k is the width or the height of the image.

Figure 4: Example of splitting the digit using a uniform grid (3x2)

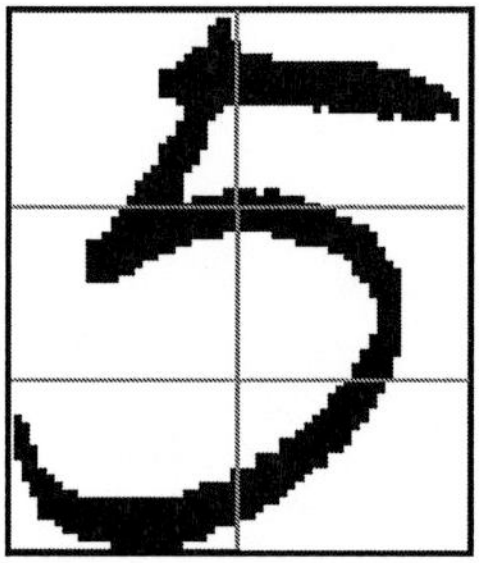

Structural Feature Generation Methods

The aim of feature generation in the recognition field is to express the features under a numerical form or symbolic system called coding (Cai and Liu 2001, Heutte et al. 1998) in order to select the best feature vector. In order to describe the main methods and to synthesize the properties which seem interesting to us, we recall the best ones best adapted in our case (Aubertet and Kornprobst 2006).

Global Features

The global features used for generating the feature vector are defined as follows:

- Density: the number of black pixels in an image divided by its size;
- Center of gravity: two coordinates to represent the gravity center of an image. They are bound by its size;
- Geometrical moment of order 2: It is a statistical measure of the allocation of pixels around the gravity center in an image bound by its size;
- Number of transitions (four main directions): It defines the number of transitions calculated between the pixels (white/black) or (black/white) normalized by the image size.

Background Features

The background features are based on information from concavity. These features arise from the topological and geometrical properties of the character classes. Each concavity feature is the number of white pixels that belong to a specific concavity configuration (Cavalin et al. 2006). The label for each white pixel is chosen based on Freeman code according to the direction. Each direction is explored until the meeting of a black pixel or limits imposed by the extremity of digits.

Figure 2: Different configurations of concavity

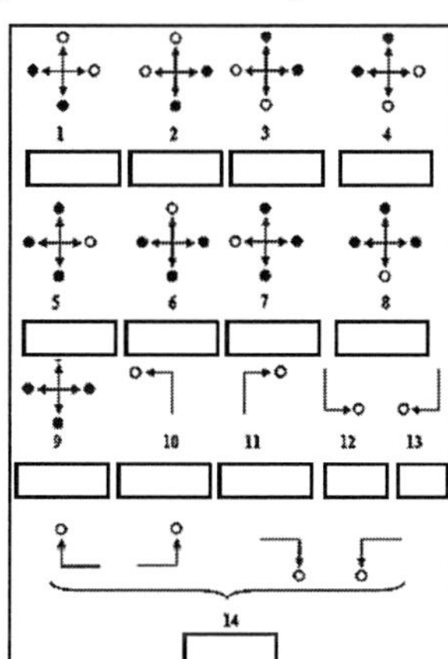

We have 9 possible concavity configurations. Moreover, we consider five additional configurations to more accurately detect the presence of loops (see Figure 3). The length of this vector is 14. The concavity vector is normalized between 0 and 1. Figure 2 shows the 9 concavity configurations and 5 configurations (10, 11, 12, 13, and 14) for the false loops.

Figure 3: Labels of concavity for the digit image '9'

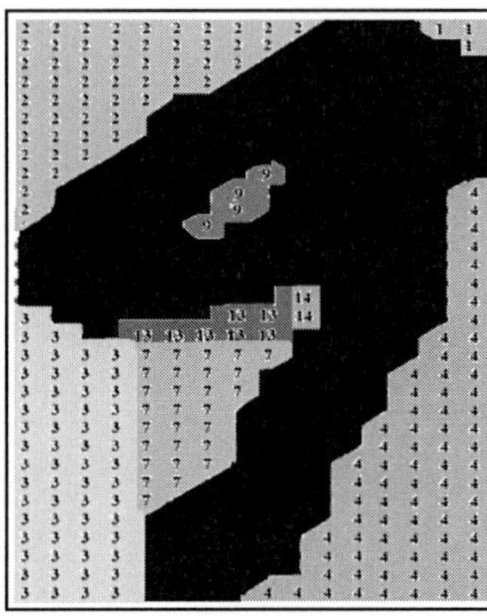

Foreground Features

This is a first step to detect contour itself, then a second step is to make the chaining (coding of contour points). There are several ways to proceed with the contour detection. Among these techniques, we use the mathematical morphology (Britto et al. 2004, Chen and Wang 2000, Oliveira 2003, Cheriet et al. 2004). Figure 4A shows how the contours are detected. The chaining process for its part is often based on the method of Freeman code applied to a string of pixels representing the contours of a digit. This code is obtained by reading the local structure of contour (string "closed" of a pixel thick).

In this method, the contour and neighboring pixels are synthesized on Freeman 8-direction (Figure 4C). For each pixel belonging to the contour, the neighborhood of a current pixel is first calculated by the 8 directions in the clockwise direction and grouped into a vector (Figure 4B). Therefore, the feature vector is composed of 8 components normalized between 0 and 1. Finally, the last part of the feature vector is bound to the surface of a digit.

Figure 4: Contour detection: (A) Contour of the upper region, (B) Feature vector, and (C) 8-Freeman directions

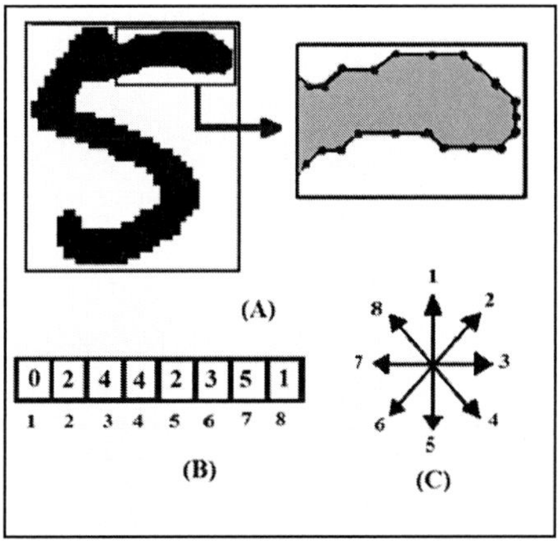

However, they are very sensitive to noise and also invariant to scale changes. Furthermore, they do not describe correctly the form of structures, such as junctions, and are poorly informative towards the image topology.

Foreground Features of the Skeleton

This method implements the skeletonization (Jang and Chin 1992, Aubertet and Kornprobst 2006). It does not allow itself to get significant features of digits. So as for the contour detection, it is necessary to complete the process by a second extraction step in order to construct a skeleton vector. This vector is built from three items: the endpoints, crossings and directional points.

To reduce the confusion between digit classes, for instance "4" and "7" (Oliveira 2003), we use feature points of the skeleton. Four components are considered: endpoints, crossing points, and two directional points that are detected when the skeletal path changes its direction to horizontal or vertical.

These points are shown in Figure 5 by the numbers 1, 2, 3 and 4, respectively. Once the skeleton points are detected, the points of each configuration are counted, normalized and used by updating the feature vector. This method is increasingly not adapted to the analysis of full forms and sensitive to noise.

Figure 5: Features generated from the skeleton

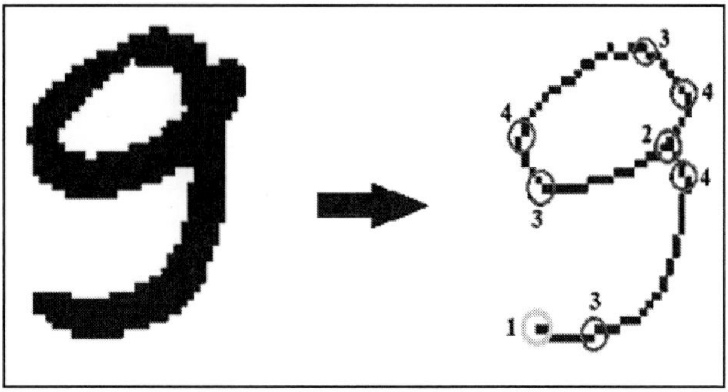

Ridgelet Transform

The Ridgelet transform has been defined by Candès and Donoho (1999) in order to improve the characterization of line singularities contained into an image (Chen et al. 2006). The Ridgelet transform combines the Radon and wavelet transforms. The Radon transform has the ability to detect lines from the image while the wavelet transform allows detection of the line singularities along the Radon slices.

Ridgelet transform was widely used with success in various applications such as image compression (Santosh et al. 2012, Yu 2012), image transform coding (Zhenghua et al. 2012), handwritten character recognition (Nemmour and Chibani 2011, Naveena and Aradhya 2011), watermarking (Mangaiyarkarasi and Arulselvi 2011), texture image retrieval (Arivazhagan et al. 2006, Mustaffa et al. 2010) and biometric identification (Arivazhagan et al. 2011).

The Ridgelet transform is based on the Radon transform which is computed according to several angular directions. Radon coefficients correspond to projections representing the shadow of the shape at each angle (Candès and Donoho 1999). Consequently, significant linear features in any direction are expressed by high magnitudes. Thus, in order to characterize linear singularities, the one-dimensional wavelet transform is applied on Radon slices to yield the Ridgelet coefficients. Hence, along the Radon axis projection, the Ridgelet is constant while in the orthogonal direction to these ridges it is a wavelet (Arivazhagan et al. 2010).

For an image $I(p_1, p_2)$, the Ridgelet transform can be computed according to the following steps:

1) Compute the Radon transform such as (Chen et al. 2006):

$$T_{rad}(\theta,r) = \iint I(p_1, p_2)\delta(p_1 \cos(\theta) + p_2 \sin(\theta) - r)dp_1 dp_2$$

δ, θ and r are Dirac distribution, angular and radial variables, respectively.

2) Apply the 1D wavelet transform on each Radon slice in order to obtain the Ridgelet coefficients $T_{rad}(\theta,r)$.

In our context, features are generated by partitioning the digit image in different blocks having the same size. Hence, the Ridgelet transform is applied on each block from which the local energy is deduced on each one.

System Evaluation

This work is devoted to generating pertinent features in order to find the best ones of normalized and not normalized isolated handwritten digits. The algorithm that we propose includes global features, the Ridgelet transform, background and foreground features (contour, skeleton), completing the system by a multi-class SVM classifier based on the One-Against-All implementation.

Databases

The NIST SD19 database is provided by the National Institute of Standard and Technology (NIST) (Grother 1995). The experiments are conducted on this database containing not-normalized isolated handwritten digits stored as binary images. The isolated handwritten digits have the advantage of being abundantly available in the form of a labeled database. This database is divided into 3 divisions (hsf_0123, and hsf_4 hsf_7). The split of each division is illustrated in Table 1.

Table 1: Distribution of the NIST SD19 database for isolated digits

Division Name	Number of images
hsf_0123	195 000
hsf_4	58 646
hsf_7	60 089

We evaluate the efficiency of various combination features on division hsf_0123 from which we extract 120,000 handwritten digits. We divide this database into three different parts: a training database containing 60,000 digit images and a validation database containing 30,000 digit images, while the remaining 30,000 digit images are kept for testing.

Classification Method

The recognition module is based on a SVM classifier, which is designed for a binary problem (two classes). To extend the binary SVM to a multi-class problem, some researchers try to make various implementations (Guermeur et al. 2000, Weston and Watkins 1999). The multiclass problem is usually treated with a combination of binary classifiers. For a problem with *N* classes, there are two types of combination methods One-Against-All and One-Against-One (Hsu and Lin 2002).

The recognition module in our system is based on the One-Against-All implementation. Two parameters should be tuned, which are σ for the Radial Basis Function (RBF) kernel and the regularization parameter *C* for the SVMs. Thus, some main properties should be considered between C and σ.

A low value of C (<1) error rate increases the value of σ.

Large values of *C* counter-balance the bias introduced by large σ.

Very small value of σ may result in good training data error rate but won't be useful in the case of test data recognition error rate.

With large value of σ, the RBF kernel becomes almost linear.

We consider an optimal pair of σ_j and C_i by testing and comparing the accuracy rates of the corresponding experimental SVMs using the validation database.

Finally, the recognition module is based on the SVM multi-class approach using the One-Against-All implementation (Guermeur et al. 2000, Hsu and Lin 2002, Weston and Watkins 1999). SVM and RBF kernel parameters are fixed to C=10 and σ= 8, respectively.

Normalization

To evaluate our technique on the same database but on the normalized digit image, we add a normalization module which allows adjusting the size, position and shape of the digit image in order to reduce the shape variation between images. In this way, the feature generation task is facilitated and to expect improving their classification (Chen and Wang 2000).

In our case, the method based on bilinear interpolation is selected which allows assigning each target point to a linear combination of four nearest source points and then performing the inverse transformation. For example, suppose that the gray levels *u(i1, i2)* of an image are known points with integer coordinates (0, 0), (1, 0), (0, 1), (1, 1). Assigning a gray scale value at the point (x, y) by a bilinear interpolation (Aubertet and Kornprobst 2006), is defined by:

$$u(x,y) = (1-x)(1-y)u(0,0) + (1-x)yu(0,1) + x(1-y)u(1,0) + xyu(1,1)$$

Other types of interpolation can be defined by using more points with integer coordinates. By these methods, it is possible to increase the resolution of images fictively. In our work, the image size is normalized to the size 24 x 24 pixels.

Experimental Results

Various feature methods are generated from the normalized and not normalized isolated digit image in order to evaluate its influence.

We test five feature sets independently to SVM classifier, which are global features, the Ridgelet transform, background features and foreground features (contour, skeleton) extracted from the images. The experiments are carried out with and/or without the uniform grid.

Experiment 1

This experiment is performed with the combination of structural features without uniform grid, in order to assess the quality of recognition. This combination appears as follows:

- Global features (8 components);
- Foreground features (8 components);
- Background features (14 components);
- Foreground features of the skeleton (4 components);
- Ridgelet transform (1 component).

Finally, the final feature vector is composed of 35 components. This evaluation is illustrated in Table 2.

Experiment 2

This experiment is performed with the uniform grid which is divided into four regions each of which is the same size. Features are generated from each region as described in Experiment 1. Finally, the global feature vector is composed of 140 (35x4) components. This evaluation is illustrated in Table 2.

Experiment 3

This experiment is conducted by combining two structural features -- background features (14 components) and foreground features (4 components) of the skeleton. For the three remaining methods (global features, Ridgelet transform and foreground features), the image is divided into four regions by using the uniform grid method. Then, we generate for each region 17 components. Finally, the global feature vector is composed of 86 (17x4+14+4) components. This evaluation is reported in Table 2.

We can note that the overall rate is acceptable for Experiment 3. However, some digits are confused and often they are visually very similar.

This evaluation is significant because the results given in Table 2 show that the combination of features results in an important improvement overall rate. However, we can see that the foreground features of the skeleton and the background features without uniform grid method allow better discrimination of digits compared to Experiments 1 and 2, respectively.

Experiment 3 shows that the use of background features without uniform grid method allows locating effectively concave regions, for example, loop, the same way as for foreground features of the skeleton. It is used in order to reduce confusion between the classes “4” and “7” for example. However, we can note that the methods of structural feature generation used work well with the database of the not normalized digits.

The digit image is divided into separate regions via splitting sampling allowing the generation of features having pertinent information. The use of a uniform grid allows this information to be improved upon for each feature by defining different levels of granularity. Features generated in various experiments show that the most effective features are structural, strokes and concavities when used in combination with a 2×2 region.

Table 2: Recognition rates for various experiments

Class	Experiment 1		Experiment 2		Experiment 3	
	Normalized	*Not normalized*	*Normalized*	*Not normalized*	*Normalized*	*Not normalized*
0	98.30	99.30	95.40	98.90	98.30	**99.60**
1	97.10	98.90	97.50	98.20	**99.20**	99.10
2	97.40	97.50	96.20	98.10	99.00	**99.40**
3	98.70	99.30	98.50	98.80	98.80	**99.10**
4	98.20	98.00	98.20	98.50	98.90	**99.30**
5	98.70	98.90	97.50	98.20	**99.50**	99.30
6	98.20	98.10	95.10	96.10	**99.20**	98.90
7	97.20	96.80	96.60	97.00	97.80	**97.80**
8	97.90	98.20	96.00	96.80	99.20	**99.20**
9	98.40	98.60	98.60	98.40	99.00	**99.60**
Over-all rate	98.01	98.36	96.96	97.90	**98.89**	**99.07**

Conclusion and Future Work

The objective of this chapter is to compare various combination features generated from not-normalized and normalized isolated handwritten digits. The combination of five feature generation methods allows a significant improvement of the recognition rate. However, we can note that the foreground features of the skeleton and the background features without uniform grid method allow better discrimination of digits. The combination of these methods can improve performance and overcome almost all problems.

Generally, the first results obtained are encouraging since we succeded with a weak recognition error. The advantage of this combination is to provide better discrimination of the digits in most cases. In future research, we will try to improve the overall recognition rate of the system by adding other feature generation methods.

References

Arivazhagan S., Priyadharshini S. S. and Sekar, J. R. (2011) "Iris recognition using Ridgelet transform", *International Conference on Recent Advancements in Electrical, Electronics and Control Engineering (ICONRAEeCE)*, pp. 286–290.

Arivazhagan, S., Ganesan, L. and Subash Kumar, T. G. (2006) "Texture classification using Ridgelet transform", *Pattern Recognition Letters*, Vol. 27, pp. 1875-1883.

Arivazhagan, S., Gowri, K. and Ganesan, L. (2010) "Rotation and Scale-Invariant Texture Classification using Log-Polar and Ridgelet Transform". *Journal of Pattern Recognition Research*. Vol. 1, pp. 131–139.

Aubertet, G. and Kornprobst, P. (2006) "Traitement des images numériques", In J. Akoka and I. Comyn-Wattiau, editors, *Encyclopédie de l'informatique et des systèmes d'information*, Vuibert, vol 18, No. 6, pp. 861–879.

Bailey, R. R. and Srinath, M. (1996) "Orthogonal moment features for use with parametricand non-parametric classifiers", *IEEE Trans. on Pattern Analysis and Machine Intelligence*, Vol. 18, No. 4, pp. 389–399.

Bortolozzi, F., Souza, A., Oliveira, L. S. and Morita, M., (2005) "Recent Advances in Handwriting Recognition", In *Document Analysis*, edited by Umapada Pal, Swapan K. Parui, Bidyut B. Chaudhuri, Kolkata: Indian Statistical Institute, pp.1–30.

Britto, A., Sabourin, R., Bortolozzi, F. and Suen, C. (2004) "Foreground and Background Information in an HMM-based Method for Recognition of Isolated Characters and Numeral Strings", *International Workshop on Frontiers in Handwriting Recognition.*, pp. 371–376.

Cai, J. and Liu, Z. Q. (1999) "Integration of structural and statistical information for unconstrained handwritten numeral recognition", *IEEE Trans. on Pattern Analysis and Machine Intelligence*, Vol. 21, No. 3, pp. 263–270.

Candès, E. J. and Donoho, D. L. (1999) "Ridgelets: A Key to Higher-dimensional Intermittency", *Philosophical Transactions of the Royal Society A: Mathematical, Physical and Engineering Sciences*, Vol. 357, No. 1760, pp. 2495–2509.

Cavalin, P.R., Britto, A., Bortolozzi, F., Sabourin, R. and Oliveira, L. (2006) "An implicit segmentation-based method for recognition of handwritten strings of characters", *in Proceedings of the 21th Annual ACM Symposium on Applied computing*, Dijon, France, April 23-27 pp. 836–840.

Chen, G. Y., Bui, T. D. and Krzyzak, A. (2006) "Rotation invariant feature extraction using Ridgelet and Fourier transforms", *Journal of Pattern Analysis and Application*, Vol. 9, pp.83–93.

Chen, Y. K. and Wang, J. F. (2000) "Segmentation of Single- or Multiple-Touching Handwritten Numeral String Using Background and Foreground Analysis", *IEEE Transactions on Pattern Analysis Machine Intelligence*, Vol. 22, no. 11, pp. 1304–1317.

Cheriet, M., Kharma, N. Liu, C.L. and Suen, C.Y. (2004.) "Character Recognition Systems, a Textbook for Students and Practitioners", *Edition John Wiley & Sons Inc*. pp.326.

Devijver P.A, and Kittler, J. (1982) "Pattern recognition, a statistical approach", *Prentice Hall*, London, pp. 480.

Favata, J. and Srikantan, G. (1996) "A Multiple Feature/Resolution Approach ToHandprinted Digit and Character Recognition", *International journal of imaging systems and technology*, Vol. 7, no. 4, pp. 304–311.

Grother, P. J. (1995) "NIST Special Database 19 – Handprinted Forms and Characters Database", National Institute of Standards and Technology (NIST).

Guermeur, Y., Elisee, A. and PaugamMoisy, H. (2000) "A new multiclass SVM based on a uniform convergence result", *IJCNN*, Vol. IV, pp. 183–188.

Heutte, L., Paquet, T., Moreau, J. V., Lecourtier, Y. and Olivier, C. (1998) "A structural/statistical feature based vector for handwritten character recognition", *Pattern Recognition Letters*, Vol. 19, No. 7, pp.629–641.

Hsu, C. and Lin, C., (2002) "A comparison of methods for multi-class support vector machines", *IEEE Trans. on Neural Networks*, Vol. 13, pp. 415–425.

Hu, J. and Yan, Y. (1998) "Structural primitive extraction and coding for handwritten numeral recognition", *Pattern Recogniton*, Vol. 31, pp. 493–509.

Jang, B. K. and Chin, R. T. (1992) "One-pass parallel thinning: Analysis, properties, and quantitative evaluation", *IEEE Transactions on Pattern Analysis and Machine Intelligence*, Vol.14, No. 11, pp.1129–1140.

Kim, K. K., Kim, J. H. and Suen, C. Y. (2002) "Segmentation-based recognition of handwritten touching pairs of digits using structural features", *Pattern Recognition*, Vol. 23, No. 1, pp.13–24.

Mangaiyarkarasi, P. and Arulselvi, S. (2011) "A new digital image watermarking based on Finite Ridgelet Transform and extraction using ICA", *ICETECT*, pp. 837–841.

Mustaffa, M. R., Ahmad, F., Mahmod, R. and Doraisamy, S., (2010) "Invariant Generalised Ridgelet-Fourier for shape-based image retrieval", *CAMP*, pp. 79–84.

Naveena, C. and Aradhya, V. N. M. (2011) "An impact of ridgelet transform in handwritten recognition: A study on very large dataset of Kannada script", *WICT*, pp. 618–621.

Nemmour, H. and Chibani, Y. (2011) "Handwritten Arabic recognition based on Ridgelet transform and support vector machines", *International High Performance and Computer Symposium HPCS'2011*, pp. 357–361.

Oliveira, L. S. (2003) "Automatic recognition of handwritten numerical strings", *Thesis of PhD*, Ecole de technologie superieure, University of Quebec. Canada, pp. 56–58, 184.

Pittner, S. and Kamarthi, S. V. (1999) "Feature extraction from wavelet coefficients for pattern recognition tasks", *IEEE Trans. on Pattern Analysis and Machine Intelligence*, Vol. 21, No. 1, pp.83–88.

Santosh, D. H. H., Dasari, P., Kiran N. L. S. and Prasad, P. S. (2012) "FRCT based efficient image compression for texture images", *IEEE ICCCA*, pp.1–6.

Weston, J. and Watkins, C. (1999) "Multi-class support vector machines for multi-class pattern recognition", *ESANN*, pp. 219–224.

Yan, D. Y. H. (2001) "Separation of touching handwritten multi-numeral strings based on morphological structural features", *Pattern Recognition*, Vol. 34, No. 3, pp.587–599.

Yu, Y.H. (2012) "A method of embedding zero-tree static image compressing based on Ridgelet transform", ICCSIP, pp. 444–48.

Zhenghua, S., Guodong L., Qing Xie, X., Zhong, R., Lvming, Z. and Lixin, G. (2012) "Orthonormal finite ridgelet transform coding using an EBCOT-like algorithm", *ICIA*, pp. 497–500.

CHAPTER NINE

TOWARD A GENERIC MODEL OF MACHINE LEARNING ALGORITHMS

GHERGHOUT YOUCEF AND TLILI YAMINA

Machine learning has become one of the major preoccupations of the computer science and artificial intelligence. (Bishop. 2006). Members of commonality learning face the significant challenge of giving machines the ability to learn and generalize.

However, these algorithms are participating in its structure to organize data, select and change the type of kernel function implemented. First, the feature selection can retain only the relevant attributes, such as maintaining performance (Blum et al., 1997, Jain et al. 1997). On the other hand, the choice of the kernel function and parameter settings affects this performance. (Schölkopf et al. 2001)

If we take a holistic view, we note that the computational implementation of these algorithms, the machine learning approach, is based on the network structure or a special case of forest (Zuluaga et al. 2010, Breiman. 2001).

The purpose of this article is to provide a rigid theoretical basis. This allows engaging in the design and the implementation of a general model, such as all of the algorithms inspired by this model.

We began our study in this context, and we were faced with a conundrum in respect to the wide range of machine learning techniques proposed in the literature. How does one create a generic structure for modeling a learning algorithm inspired by the queuing system to implement a dedicated system, for example the classification of medical images? This system allows the aggregation of the features or instances of learning/test that are out of the original set. We must recognize the importance of this model through the use of these systems and be able to solve these types of problems.

In this article, we describe the architecture of the most popular algorithms of machine learning. We then propose a generic structure based on the queuing network model to model and by the classifiers generated by the machine learning algorithms. The performances are in terms of recognition rate, error rate, and the cost of computation time retained by the processor.

Modeling Techniques of Machine Learning Algorithms

Machine learning concepts encompass any method that builds a model of reality from data. There are two major trends in learning, the issue of artificial intelligence called symbolic intelligence, and the outcome of statistics and qualified digital.

Artificial neural networks

Dreyfus (2005) generates a formal description of the real world system through the conceptualization of the neural network. It performs a thorough treatment of the principles of modelling to provide a comprehensive methodology for designing a model.

The proposed model is matched with the different classes of the neural network, either Feedforward Neural Network or Recurrent (Feedback) Neural Network, and either "black-box" or "gray-box" models.

The Queuing Network model organizes the features of the input instances and is ranked according to their pertinence. It facilitates the classification, discrimination, and prediction of the input data.

Support vector machines

The SVM is a discriminative model that attempts to minimize learning errors while maximizing the margin between data classes. Maximizing the margin is a regularization method which reduces the complexity of the classifier. The penalization of SVM parameters in order to maximize the margin is a method of selecting an implicit model for learning. This process generates a reduced set of prototypes. This part of the training set is commonly called support vectors. Its behaviour is also conditioned by the type of kernel used and the values granted to its parameters. The kernel of an SVM is a symmetrical, positive-defined function that projects the data in a transformed high-dimensional space. The separation of classes easily operates in this space, in the same way that the hidden neurons of a PMC project the data entered into a representational space in which the weight of the output layer of classes defines linear discriminated function. The values of the kernel parameters directly affect the complexity of the decision boundary of the classifier. The values also influence the number of support vectors of the classifier. Fig. 1 and Table 1 show respectively the architecture of SVMs and some kernel functions, respectively. This kernel function k is chosen a priori; it determines the type of classifier, e.g. polynomial classifier, radial basis function classifier, or neural network. All other parameters (number of hidden units, weights, threshold b) are found during training, by solving a quadratic programming problem. The first layer weights xi are a subset of the training set (the Support Vectors); the second layer weights $\lambda_i = y_i * \alpha_i$, are computed from the Lagrange multipliers (Schölkopf et al. 2001).

Figure: 1 Architecture of SVMs

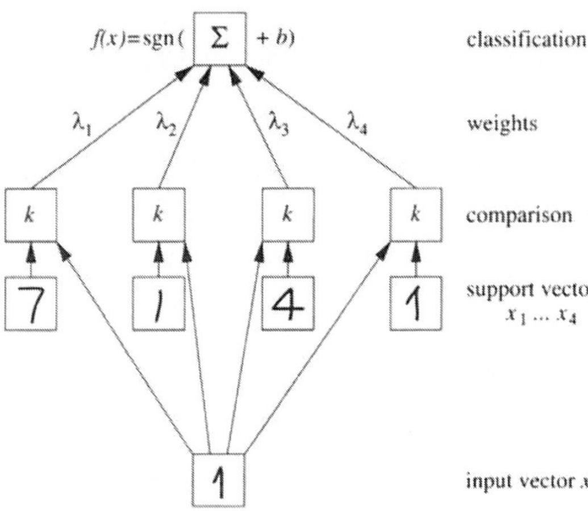

Table 1: Some Classical Kernel Functions

Classification: $f(x) = \text{sgn}(\sum_i \lambda_i k(x, x_i) + b)$

Weights: $\lambda_1, \lambda_2, \lambda_3, \lambda_4$

Comparison	
Classical architectures: *polynomial classifiers* of degree *d*	$k(x, x_i) = (x.x_i)^d$

Radial basis function classifiers with Gaussian kernel of width *c> 0,*	$k(x,x_i)=\exp(-\|x-x_i\|^2/c)$
Neural networks with tanh activation function	$k(x,x_i)=\tanh(k(x,x_i)+\theta)$
Support vectors:	$x_1,\ldots,x_4$
Input vector:	x

Decision trees and rule learners

Data mining supports the complete process of knowledge extraction, such as storage in a database and data selection to study. If necessary, data cleaning then uses the numeric and symbolic learning to propose models to user validation of the proposed models. Finally, if these models are invalidated by the user, the entire process is repeated.

There are various descendent decision trees learners (Breiman et al. 1984):

- ID3 "Decision Tree Induction" is the designation of an old building algorithm graph tree induction, in which the criterion variable selection is informational gain. ID3 is the most basic strategy to build decision trees.
- C4.5, which is the descendant of ID3, is a whole program for building tree induction using a criterion of a more elaborate "gain ratio" and to simplify rules for handling missing data, etc… (Breiman et al. 1984).
- CART (classification and regression trees) is proposed by a general methodological framework to generate serious binary induction. It is not, strictly speaking, a specific method. From this point of view, CART is similar to the general segmentation (Breiman et al. 1984).
- Random forest (RF) is a classification method that affects some objects into one of the given categories and provides feature importance. For a specific pattern classification problem, a random forest is computer software that is developed so that objects are classified correctly with reasonably good performance. The classifier of the Random Forest is composed in its turn of a collection of tree-structured under-classifiers. Each of these is

constructed by instances with randomly sampled features. Predictions are made by a majority vote of the trees. A study on Random Forest is presented by Breiman in his article (Breiman. 2001)

To formulate the rules of the tree, each branch of the tree represents a profile that is in the language, a combination of proposal types (attribute-value) that can write to the formalism in predicate logic without difficulty.

k – Nearest neighbor techniques

Nearest neighbor techniques are based on an assumption about the world: nearby points are more likely to be given the same classification than distant ones. Suppose we have a set of points in a stated space that has been given classifications. Suppose a new pattern is to be classified. The distance from the new pattern to all the old ones is computed. The new pattern is given the classification of the previously classified point that is closest to it, that is, its nearest neighbor (see Fig.2) (Anderson. 1997).

One version of a nearest neighbor classifier can be realized easily using the generic connectionist neuron. Neural network model neurons, the generic computing units, can compute comparative distances between examples because of the close relation between the inner product of the weights and the input pattern and Euclidean distance. Figure 3 shows the simple nearest neighbor classifier, where the distances from a new point to the classified old ones are computed. The new point is given the classification of the nearest old point.

Figure 2: A simple nearest neighbor classifier.

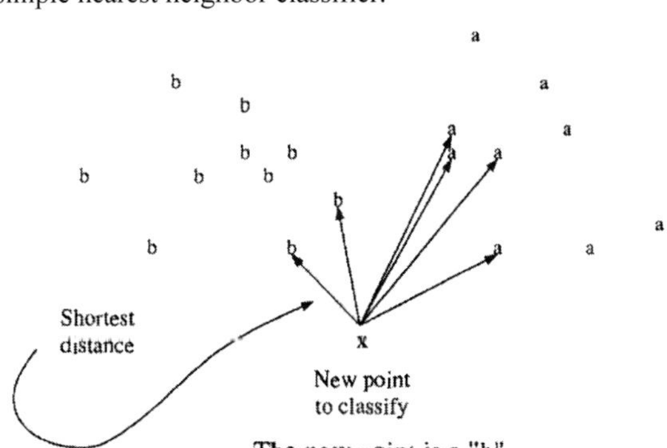

If we want to use more complex nonlinear units, this technique will still work if the nonlinearity in the second stage is monotonic and the same for all units. The inner product is what provides the selectivity for the system. The weights at the middle layer units reproduce the values of a classified input pattern. The length of all the input patterns is the same. If a new pattern is presented to the network, the classification of the most active unit will be the nearest neighbor classification.

Figure 3: A slightly nonstandard neural network

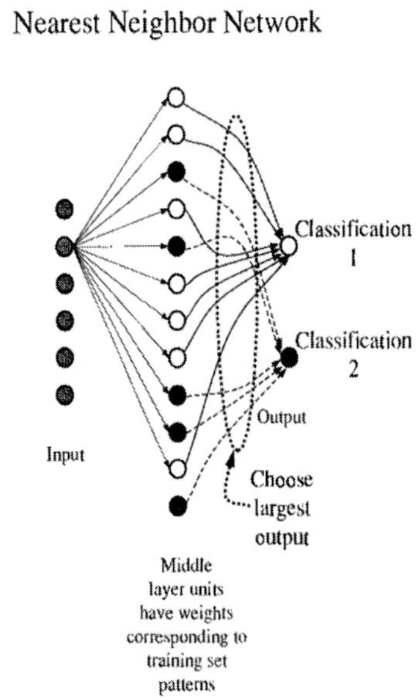

Architecture of the Proposed Model

The architecture of our proposed model contains three basic components:

Performance estimation and Kernel function choosing

The performance is measured after the training takes place on a certain number of data aggregated, which is called process of training. These data, on their turns, are described together by an aggregated feature set. However, it is necessary to ensure that the measure of performance takes place on the processes of the test that differ from the training processes.

The measure of performance is controlled by the parameter of the kernel function chosen by the user, and this is in terms of the classification rate and the computational time retained by the processor.

Choosing the type of kernel function and its degree of complexity differs from a classification/prediction problem to another, a deviation of the kernel function affects the results generated by classifiers. Its parameter does not reflect or correspond to the optimal solution, and this leads to a relative decrease in performance. This is chosen a priori by the user, who determines the type of classifier. Other parameters (number of hidden units, weights, threshold b) are found during the learning. On the other hand, the degree of the relevance of the feature differs from one characteristic to another; some characteristics are stronger and more descriptive than others. This is what motivated us to achieve the techniques for selecting these features. This procedure is according to the selection and ranking of these features, depending on the degree of its strengths and its relevance in the system of classification.

At the time of the selection of every feature, an aggregation to the level of the training instances that only represents the selected features is calculated. The time of CPU processor is calculated before and after the process of selection (See Fig. 4).

The Modeling (Classifier/Predictor)

The result of training is also called a classifier model of identification, pattern recognition, predictor, etc. The relevance of a classifier can be appreciated in different ways. One agrees to recognize the importance of some global criteria that we clarify in this paragraph (Zuluaga et al. 2010, Gangeh et al. 2010):

- Precision: It is certainly the most important criteria, and in all cases, this is the most often quoted of the supervised training. It shows the intrinsic capacity of the classifier to recognize the variable to predict in the population, when the precision is perfected.
- Simplicity: It is often bound to the number of parameters that intervene in the model. The model is much more intelligible less parameters it has.
- Comprehensibility: The extracted knowledge must be semantically and structurally similar to those that a human expert can produce.
- Intelligibility: The engineering solution brought by the KDD (Knowledge Discovery in Databases) is bound to the intelligibility

of the model constructed automatically. A system of excavation of data is a product of the knowledge that can be under varied shapes according to the method of training used: of the decision trees, of the rules of associations, etc.

Figure 4: Proposed System Architecture: Modeling of the machine learning algorithms

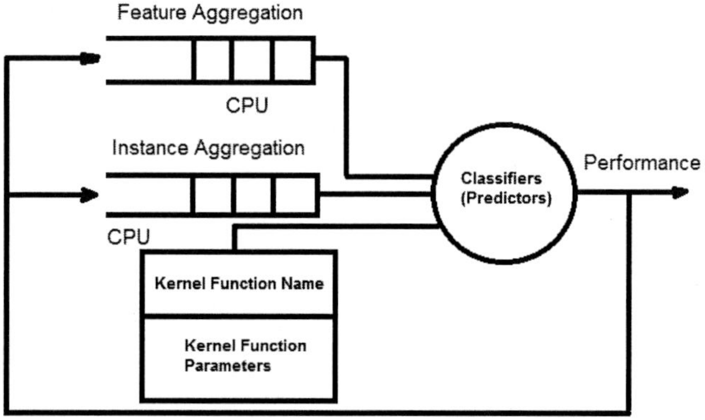

Conclusion

In this study, we have demonstrated the need for a generic structure based on the modeling of Queuing Network System to aggregate features and instances learning simultaneously. Both sets obtained are injected into learning process multi classifiers/predictors. Thus, the parameters of the kernel function are provided by the user and set it to the performance of a given classification.

References

Anderson, J. A. (1995) *An Introduction to Neural Networks*, Cambridge, MA: The MIT Press.

Bishop, C.M. (2006) P*attern Recognition and Machine Learning*, Information Science and Statistics, New York: Springer-Verlag, Inc.

Blum, A. L., Langley, P. (1997) "Selection of Relevant Features and Examples in Machine Learning", *Journal of Artificial Intelligence,* Vol. 97, No. 1 & 2, pp. 245-271.

Breiman L. (2001) "Random forests", In: *Machine Learning, Statistics Department,* Vol. 45, pp. 5-32.

Breiman, L., Friedman, J., Olshen, R. and Stone, C. (1984) "Classification and Regression Trees", pp. 81-106, Chapman and Hall/CRC.

Dreyfus, G. (2005) *Neural Networks Methodology and Applications*, Springer, pp. 1-497.

Gangeh, M. J., Sørensen, L., Shaker, S. B., Kamel, M. S., and Bruijne, M. (2010) "Multiple Classifier Systems in Texton-Based Approach for the Classification of CT Images of Lung". *Proceedings of the 2010 international MICCAI conference on Medical computer vision: recognition techniques and applications in medical imaging,* pp. 153-163. Springer-Verlag.

Jain, A., Zongker, D. (1997) "Feature Selection: Evaluation, application, and small sample performance", *Journal of IEEE Transactions on Pattern Analysis and Machine Intelligence,* Vol. 19, No. 1 & 2, pp. 153-158.

Schölkopf, B., Smola, A. J. (2001) *Learning with Kernels: Support Vector Machines, Regularization, Optimization, and Beyond*, Cambridge, Massachusetts: The MIT Press.

Zuluaga, Maria A. Leyton, Edgar J. F. Delgado, Marcela Hern´andez, and Orkisz Maciej (2011) "Feature Selection for SVM-Based Vascular Anomaly Detection", *Proceedings of the 2010 international MICCAI conference on Medical computer vision: recognition techniques and applications in medical imaging,* pp. 141-152. Springer-Verlag.

CHAPTER TEN

A GENETIC ALGORITHM APPROACH TO QOS-BASED SEMANTIC WEB SERVICES COMPOSITION

AMINA BEKKOUCHE,
SIDI MOHAMED BENSLIMANE
AND MOHAMED MERZOUG

Web services technology offers great potential to overcome the problems of interoperability. It is a promising framework for the integration of applications and for managing interactions among various partners in a distributed environment, which is the Web.

The basic infrastructure around the Simple Object Access Protocol (SOAP) standard, Web Services Description Language (WSDL), and Universal Description, Discovery, and Integration (UDDI) is sufficient to establish interoperable and integrated components, but insufficient to make automatic and efficient multiple tasks in the lifecycle of web services, for example, the composition and also the discovery of the required services.

The need for automation of design and implementation of web services joined the concerns underlying the Semantic Web (Berners-Lee et al. 2001), specifically, how to formally describe the knowledge to make it usable by machines. The term Semantic Web Services can be found at the convergence of two important areas of research on Internet technologies, the Semantic Web and Web Services (Cardoso 2007).

Here we focus on web services composition, where a set of services is composed to satisfy the user request based on the semantic descriptions of web services and their non-functional properties (QoS parameters). The user may also express requirements through non-functional constraints (QoS constraints). To evaluate the degree of similarity between a given request and a composite service, we propose a mechanism for semantic matching based on ontological concepts annotating the input/output parameters of web services participating in the composition as well as the ontological concepts annotating the input/output parameters of the request.

The purpose of this work is to present an approach for web services composition based on a genetic algorithm that focuses on two aspects in order to find the optimal solution:

- The functional aspect: the semantic descriptions of web services are used to calculate the degree of semantic correspondence between the user request and a candidate solution (composite service).
- The non-functional aspect: the values of QoS (cost, response time, etc.) of the service components are used to calculate the total value of the solution candidate QoS.

The rest of this paper is organized as follows. Section 2 presents the proposed architecture for service composition. The semantic model and QoS model are described respectively in Section 3 and Section 4. Section 5 presents our genetic algorithm including the strategies of selection, mutation and fitness function. Section 6 shows the obtained results. Section 7 comments briefly on related work. Finally, in the last section, we present our conclusions and we comment on the direction of future work.

Related Work

A great deal of research has been done in recent years on the optimizing of web service compositions. However, no approach has addressed specifically the service composition using both QoS and semantic similarities dimensions in a context of significant scale.

In Canfora et al. (2005), an empirical study compared a GA with Integer Programming (IP)-based algorithm. His research focused on how genetic algorithms are best suited for service composition when compared to integer programming. Continuous GA has been applied to various fields of science from physics, and mathematics to medical sciences.

Lecue and Leger (2006) present a formal model for the composition of semantic web services. Semantic connections between web services are treated as causal links between operations (I/O) where they are weighted by the degrees of semantic mapping ("exact", "plug-in", "subsume", "disjunction") as introduced by Massimo et al. (2002). Their model composition using AI planning follows a backward chaining method of web services. An originality of the method lies in the composition optimization provided by a pre-calculation of semantic mappings of input and output of all discovered web services and compatible with the overall goal to achieve, in a matrix known as the "causal link" which retains only the input-output links semantically compatible. However, this method does not take into account the effects and preconditions.

In Talantikite et al. (2009), compositions are discovered within a semantic web services network by a forward chaining algorithm. Nodes represent the semantic web services and links of interaction, called "semantic dependency links" are calculated using operator mapping semantics as introduced by Massimo et al. (2002). Knowing they seek, above all, to meet the goals of the request, they propose the use of a backward chaining algorithm to pass for automatic search for compositions. One solution is selected from the set of plans compositions returned from a quality criterion.

In Thomas et al. (2007), authors propose three methods for the automatic composition of web services based on their semantic descriptions: an uninformed search in form of an IDDS (Iterative Deepening Depth-First Search) algorithm, a greedy informed search based on heuristic functions, and a genetic algorithm. The I/O parameters of web services are annotated with semantic concepts. The user needs are expressed via a query that also is characterized by the concepts of input and output concepts. To satisfy a given query, the input concepts of service (composite) should be more general than the input concepts of the query. However, the concepts of output service must be more specific than the concept of query output.

Imad (2010) proposes an approach based on a genetic algorithm for the composition of web services. The consumer expresses the web service desired (the request) by an WSDL file, and the composition system asks a search space in the form of collection of web services and returns as result a list of possible services (population).The criterion for selecting a service is its syntactic similarity value with the target service. Elements of the population result are services that exist in space research (discovery of an appropriate service) or that have been created through the composition of existing services. However, in this work only the syntactic descriptions of web services are considered and no non-functional parameters are specified.

Proposed Architecture

The aim of our system composition is to find the optimal composition best suited to the user request. It is based on a mono-objective genetic algorithm. We are interested in genetic algorithms for two reasons:

- They have proven their effectiveness in various areas to solve optimization problems and research (air traffic management, scheduling, image processing, etc.).
- They do not impose any assumption on the resolution of a given problem, and the idea behind them is to define an objective function to find a good quality solution.

Our genetic algorithm receives as input the user requirements (the Input and Output parameters as well as the global QoS constraints), and returns an optimal solution that is a composition of web services that takes into account both functional and non-functional properties of web services (See Figure 1).

The web Services are described in the SAWSDL[1] language. This language provides a mechanism for annotating semantically the services described in WSDL and XML schemas associated. These semantic annotations establish the correspondence between each element of a WSDL description (inputs, outputs, preconditions, effects) and the semantic concepts in an external domain semantic model.

The domain model consists of a domain ontology that is a hierarchical structure that allows deducing generalization relationships (subsumption) between concepts, which will be useful in determining the degree of similarity between a given request and a composite service.

In addition, an XML file representing the QoS attributes (response time, cost, availability, reputation, etc.) is associated to each web services description.

The user requirements are formulated through a web-based graphical interface. The user specifies the Input/Output parameters using the corresponding ontology concepts. Moreover, the user has the possibility of introducing global QoS constraints (e.g., on a total cost of the desired composition).

[1] www.w3.org/TR/SAWSDL/.

Figure 1: The Proposed Composition Architecture

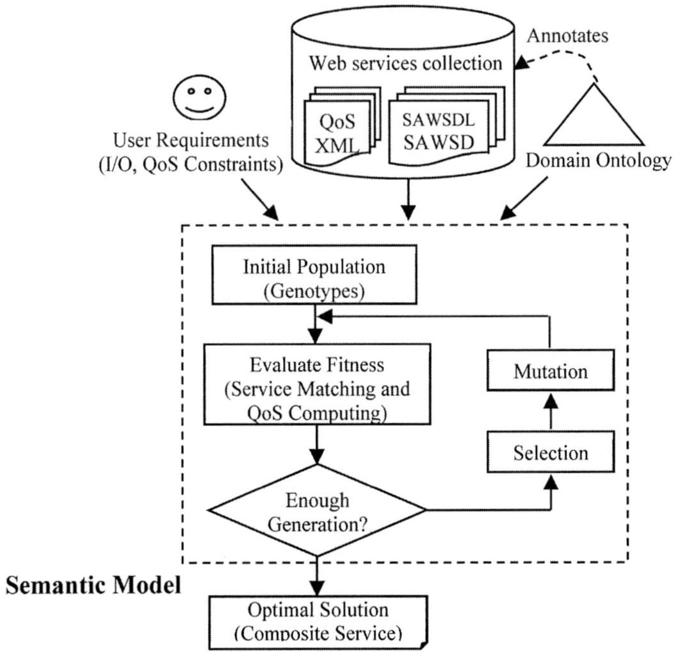

Conceptual representation

In this section we give some basic definitions of service, composite service, and request.

Definition 1 (Web Service)

A web service is described by functional and non-functional properties. Formally:$S = \{ID, IC, OC, q\}$

Where *ID* is the service identifier, *IC* and *OC* are the ontology input and the output concepts respectively, *q* is the QoS attributes list (e.g., cost, response time, availability, etc.).

Definition 2 (Composite Service)

A composite web service (CS) is represented by a set of web services. Formally: $CS = \{S_1, S_2, \ldots, S_n\}$

Where *N* is the number of web services.

Definition 3 (Request)

The user requirements are represented by a request, and defined as follows: $R = \{IC, OC, G\}$

Where *IC* and *OC* represent respectively the inputs and the outputs ontology concepts of *R*. *G* is the global QoS constraints list.

Computing The Composite Service Similarity Degree

Degrees of matching are determined in different ways, depending on the semantic descriptions of the elements to match. There are three main approaches to matching:

- *IO-matching*: also called "IO-matching service profile." This type of match is determined from semantic data service parameters: input (I) and output (O). This type of matching is adopted by (Srinivasan et al. 2004, Fan et al. 2005, Massimo et al. 2002).
- *PE-matching:* This type of match is determined from matching on pre-conditions (P) and effects (E) services and requests. This type of matching is adopted with CDR (Schumacher et al. 2008) with preconditions and effects specified in the Prolog.
- *IOPE-matching:* This type of match is determined from matching the semantic data input (I), output (O), preconditions (P) and effects (E) services and requests. This matching is adopted in (Jaeger et al. 2005, Keller et al. 2005, Küster and König-Ries 2008, Stollberg et al. 2007).

In our approach, we have adopted an IO matching based on *subsumption* reasoning to calculate the degree of similarity between the user query and the composite service.

To achieve semantic matching, we use the four kinds of matching – Exact, Plug-in, Subsume and Fail – introduced in (Massimo et al.2002), knowing that in the discovery of web services, matching is done on the same categories of parameters, i.e. the input pairs, or the output pairs. In contrast, in the composition of web services, the problem is addressed by the study of matched pairs of distinct input/output parameters of different web services of the same workflow (composition).

- *Exact:* two parameters are called similar if the concept annotating the output parameter (of the first service) and the concept annotating the input parameter (of the second service) are equivalent.

- *Plug-in:* two parameters are called similar if the concept annotating the output parameter (of the first service) is more specific than the concept annotating the input parameter (of the second service). In other words, the concept annotating the input parameter subsumes the concept annotating the output parameter.
- *Subsume:* two parameters are called similar if the concept annotating the output parameter (of the first service) is more general than the concept annotating the input parameter (of the second service).
- *Fail:* means that there is no subsumption relation between concepts.

These operators can be ordered according to a degree of relevance as follows: exact > plug-in > subsume > fail (Massimo et al.2002).

Definition 4 (Degree of Elementary Matching)

Given a composite service $CS = \{S_1, S_2, \dots, S_n\}$ and a request *R*=$\{ IC, OC, G\}$, a matching is called *elementary* when applied to the comparison of concepts pairs annotating the *I/O* parameters of the request and service components such as:

$EM_S(S_i.OC, S_{i+1}.IC)$: refers to the degree of elementary matching between the output concept of service S_i and the input concept of service S_{i+1}.

$EM_{RIN}(R.IC, S_1.IC)$: refers to eans the degree of elementary matching between the input concept of request and the input concept of service S_1.

$EM_{ROUT}(R.OC, S_N.OC)$: refers to eans the degree of elementary matching between the output concept of request and the output concept of service S_N.

The functionsEM_S, $EM_{RIN}$and EM_{OUT}return a values in [0,1] depending on the match type between the concepts. In Table 1, we summarize the elementary matching functions with their integer values.

Table 1: Elementary Matching Functions

Match Type	Exact	Plug-in	Subsume	Fail
$EM_S(S_i.OC, S_{i+1}.IC)$	1	2/3	1/3	0
Logic meaning	*S_i.OC and S_{i+1}.IC are equivalent*	*S_i.OC is subsumed by S_{i+1}.IC*	*S_i.OC subsume S_{i+1}.IC*	Otherwise
$EM_{RIN}(R.IC, S_1.IC)$	1	2/3	1/3	0
Logic meaning	*R.IC and S_1.IC are equivalent*	*R.IC is subsumed by S_1.IC*	*R.IC subsume S_1.IC*	Otherwise
$EM_{ROUT}(R.OC, S_N.OC)$	1	2/3	1/3	0
Logic meaning	*R.OC and S_N.OC are equivalent*	*R.OC is subsumed by S_N.OC*	*R.OC subsume S_N.OC*	Otherwise

Definition 5 (Degree of Global Matching)

A matching is called global when producing the final evaluation values of all matching elements. Given a composite service $CS = \{S_1, S_2, \dots, S_n\}$ and a request $R = \{IC, OC, G\}$, the degree of *global* matching or degree of composite service similarity can be calculated by:

$$GM(CS) = ((EM_{RIN}(R.IC, S_1.IC) + \sum_{i=1}^{N} EM_S(S_i.OC, S_{i+1}.IC) + \frac{EM_{ROUT}(R.OC, S_N.OC)}{(\mathrm{N}-1)+2} \quad (1)$$

Where N represents the number of component services.

QOS Model for Composite Web Service

In this section, we first present the quality criteria in the context of elementary services, before turning our attention to composite services.

Quality Criteria for Elementary Services

The non-functional properties of web services are expressed by QoS attributes such as response time, cost, reputation, performance and availability. The set of QoS attributes can be divided into two groups: positive and negative QoS attributes. The values of positive attributes need to be maximized (e.g., availability), however the value of negative attributes need to be minimized (e.g., response time) (Alrifai and Risse 2009).

We use the vector $q = \{q_1(S), q_2(S), \dots \dots \dots, q_l(S)\}$ to represent the QoS attributes of a service S, where $q_i(S)$determines the value of the ith quality attribute of S (Alrifai and Risse 2009).

Quality Criteria for Composite Service

Definition 6 (QoS of Composite Service)
The QoS for a composite service can be defined by:

$$Q(CS) = \{Q_1(CS), \dots \dots \dots, Q_l(CS)\} \quad (2)$$

Where $Q_i(CS)$represents the estimated value of the ith QoS attribute of CS.

The QoS Aggregation

The QoS of a composite service *CS* is decided by the QoS values of its component services as well as the composition model used (e.g. sequential, parallel, conditional and/or loops). In this work, we focus on the sequential composition model (in a sequence, a web service is available once the previous web services have completed their execution). Therefore, the estimated value of the QoS attribute of CS can be aggregated from the expected QoS values of its component services (Alrifai and Risse 2009).

The Global Qos Constraints

Global QoS constraints represent the user requirements in terms of QoS. They can be expressed by the lower bounds (minimum values) and upper bounds (maximum values) values of overall composite service QoS. Global QoS value constraints are constraints proposed by users, such as the service price not going beyond \$100, service availability not being under 99.9 percent, etc.

Genetic Algorithm

As described in the introduction, this work aims to propose an approach based on GA to determine an optimal composition of atomic web services. Such a composition needs to:

1. Optimize a function that calculates the degree of similarity between the user request and the composite service.

2. Optimize a function of some quality criteria.
3. Satisfy the end-to-end user's global QoS constraints (e.g, the service user may have a limited budget and thus the cost is constrained).

GA Parameters

The near-optimal solution (represented by a genotype) is determined by the evolution of an initial population through a specified number of generations by applying genetic operators (crossover, mutation…) until a stopping criterion is satisfied. In the following, we describe these parameters.

Genotype: it is defined by an array list. The number of arrays is equal to the number of web services involved in the composition, each array represent a web service that is defined by its identifier, its input concept, its output concept, and its QoS attributes. Each composition, as a potential solution of the optimization problem, can be encoded using this genotype (Lecue 2009). Figure 2 represents a genotype encoding of a composite service with three services, and two quality criteria: the response time and the cost.

Figure 2: Genotype Encoding for Service Composition

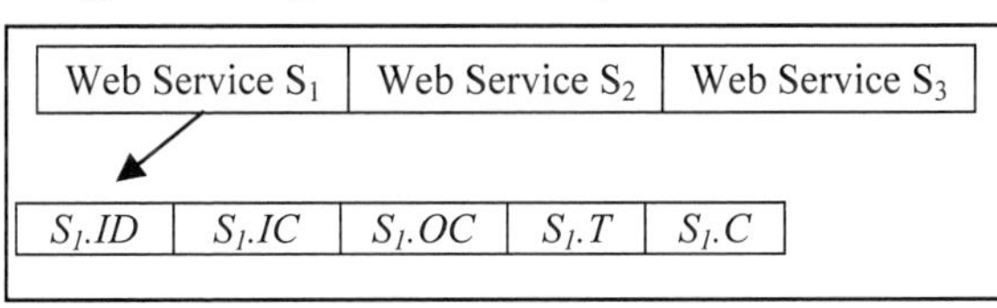

S1.ID: the identifier; S1.IC; the input concept, S1.OC; the output concept, S1.T; the response time, and S1.C; the cost.

Initial population consists of an initial set of compositions (characterized by their genotypes) wherein services are selected randomly.

Global QoS Constraints have to be met by a composite service CS e.g., 10< T (CS) <50 (where 0 and 50 are respectively the minimal and maximal value of the response time constraint).

Fitness Function is required to quantify the quality of any CS. Such a function needs:

1. To maximize the degree of similarity of CS,
2. To minimize QoS attributes of CS.

In order to obtain one function to maximize, we convert the two negative QoS attributes into positive attributes by multiplying their values by (-1):

$$F(CS) = W_s GM(CS) + \sum_{i=1}^{l} W_i Q'_i(CS) \quad (3)$$

Where Q'_i refer to Q_i (the ith QoS attribute of CS) normalized in the interval [0, 1], l is the number of QoS attributes, and $W_j \in [0, 1]$ is the weight assigned to the jth fitness factor where $\sum W_j = 1$.

In this manner, the user can adjust the weight factors W_j of the fitness function according to their preferences. Furthermore, the global constraints must be satisfied, and the compositions that do not meet the constraints are penalized.

Of the different penalty functions that are proposed in the literature (static, dynamic, adaptive functions, etc.), we have chosen the penalty function adopted by (Lecue2009).

$$F'(CS) = F(CS) - W_p \sum_{i=1}^{l} \left(\frac{\Delta Q'_i}{Q'^{max}_i(CS) - Q'^{min}_i(CS)}\right)^2 \quad (4)$$

Where Q'^{max}_i, Q'^{min}_i are respectively the maximum and minimal values of the ith QoS constraint, l is the number of QoS attributes, W_p weights the penalty factor, and $\Delta Q'_i$ is defined as follows:

$$\Delta Q'_i = \begin{cases} Q'_i - Q'^{max}_i \text{ if } Q'_i > Q'^{max}_i \\ 0 \text{ if } Q'^{min}_i \leq Q'_i \leq Q'^{max}_i \\ Q'^{min}_i - Q'_i \text{ if } Q'_i < Q'^{min}_i \end{cases} \quad (5)$$

Genetic Operators: with a not homogeneous population of individuals, the diversity of the population must be maintained over generations. This is the role of these operators. In this paper, we use the following genetic operators:

Mutation: we specified a specialized mutation operation inspired from that proposed by (Thomas et al. 2007). The operation either deletes (*mutate1*) a web service in CS or replace (*mutate2*) a web service in CS with another web service, such as the degree of composite service similarity not being null. This operation is performed in order to ensure a minimum similarity between the user request and the composite service. Unlike the situation described in (Thomas et al. 2007), the deleted and the replaced web services are chosen randomly.

We use an adjustable variable σ as a threshold for applying *mutate₁* or *mutate₂*.

$$mutate_1(CS) \equiv \begin{cases} CS' = \{s_1, s_2, \dots, s_{|CS|}\} - s_{1 \le i \le |CS|} & if\ |CS| > 1 \\ CS & otherwise \end{cases} \tag{6}$$

$$mutate_2(CS) \equiv CS' + s_{1 \le i \le |CS|} \mid GM(CS' + s_{1 \le i \le |CS|}) > 0 \tag{7}$$

$$mutate\ (CS) \equiv \begin{cases} mutate_1(CS) & if\ random() > \sigma \\ mutate_2(CS) & otherwise \end{cases} \tag{8}$$

Tournament selection: this operator is used to identify statistically the best individuals (candidate solutions) of the current population that will be allowed to reproduce, according to their fitness value. In our GA, we applied a binary tournament selection.

Stopping Criterion: determines when optimization processes carried out by Optimizers should halt. In our optimization process, we iterate until the global QoS constraints are met ($\Delta Q_i = 0$) within a maximum number of generations.

The Experimentation

To illustrate the validity and the effectiveness of the proposed approach, and to assess its performance, we implemented a java-based prototype under the operating system Windows XP and the environment eclipse with free tools and open source.

On the one hand we used several Java-API for the handling of the web services, and, on the other hand, we used Pellet Reasoner to calculate the degree of similarity (see Figure 3).

Given the unavailability of a great corpus of real services, we have performed various tests on set services we developed for the need of the application. We have incorporated estimated values for QoS parameters (cost and response time).

The adopted genetic algorithm possesses the following attributes:
Population size=50;
Mutation rate=0.05;
Genotype size (web services number)=5;
Maximum number of iterations=50.
The weights are fixed as follow: Ws=0.5, W_T=0.25, W_C=0.25,W_P=0.5.

Figure 3: The Prototype Screenshot

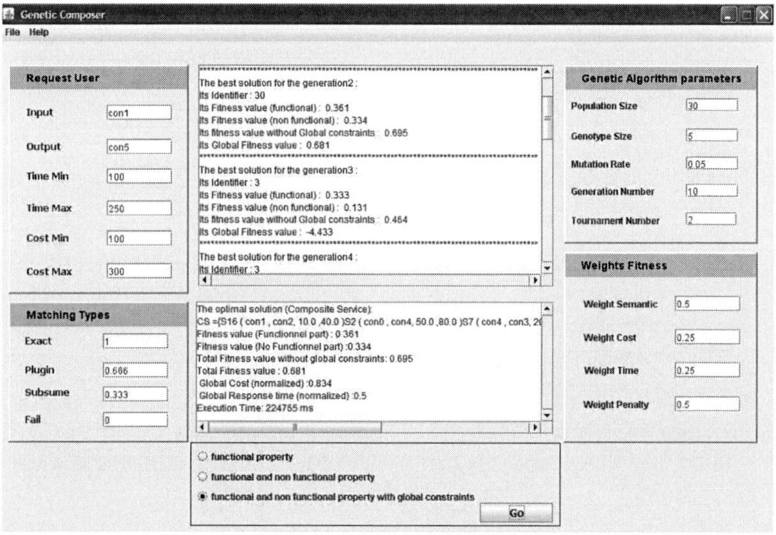

Figure 4: The Score Evolution Of The Best Configuration

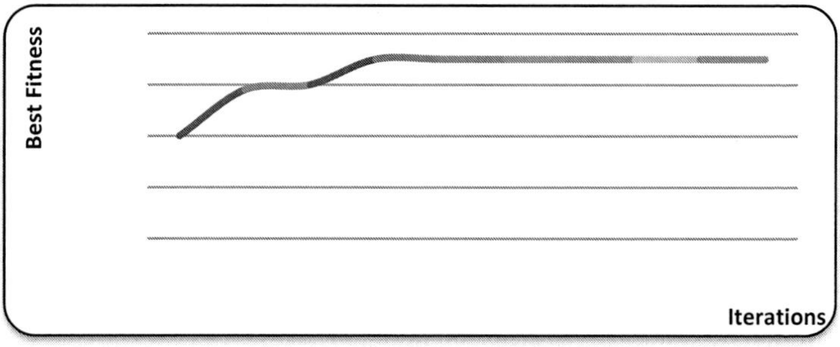

The adopted penalty function has several advantages in comparison with other propositions. It is characterized by a reduced number of parameters in comparison with other approaches. Figure 4shows the best score of the objective function that is obtained according iterations. The quick convergence to the optimal solution illustrates the applicability and efficiency of the proposed GA.

Figure 5: The Effect Of The Genotype Size On The Execution Time

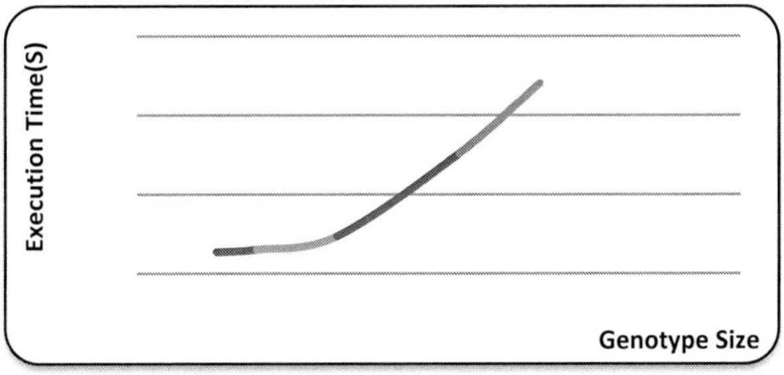

To evaluate the effect of the genotype size (service composition length) on the execution time to find the optimal solution, we have tested the performance of our GA under different genotype sizes.

The results are shown in Figure 5.We observe that as the number of services increases (genotype size) the service composition time will be increased to converge on the optimum.

In addition, another experiment was performed. We compared the best score of the objective function during the execution of the GA, taking the following into consideration: first, only the functional properties of the web services; second, functional and non-functional properties; and third, we add the global QoS constraints.

Figure 6 shows the results of this experiment by executing the GA many times. As can be seen from this figure, the GA with functional properties had a higher value of fitness function. We notice also that the GA with functional and non-functional properties gives better results in fitness value than the GA with functional and non-functional properties with global QoS constraints.

Figure 6: Evaluating the Fitness Function With Different Web Services Properties

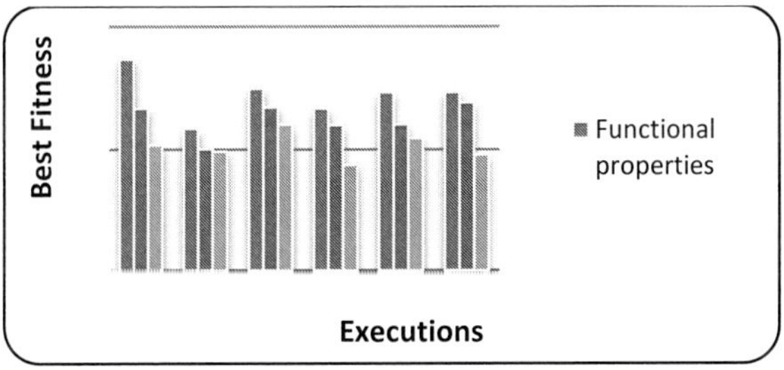

Conclusion

In this paper, we have presented a genetic-based approach for the selection of the best composition of services suited to the user requirements. The composition process takes into account both functional and non-functional properties of web services. The empirical evaluation of the proposed approach highlighted the benefits provided by the combination of semantic and QoS to enhance the service composition process.

However, it still is necessary to conduct further empirical evaluation to show that the proposed approach selects the best quality of composition services. Particularly, in the future work we will consider that web services are also characterized by their pre-conditions and effects, so that the *PE matching* or *IOPE matching* can be tested. Furthermore, we may hybridise the genetic algorithm with other meta-heuristic optimization techniques (tabu search, colony optimization, and particle swarm optimization) in order to increase its effectiveness.

References

Alrifai, M. and Risse, T. (2009) “Combining global optimization with local selection for efficient QoS-aware service composition”. In: WWW 2009: Proc. of the 18th International Conference on World Wide Web, pages 881–890.

Berners-Lee,T., Hendler, J. and Lassila, O. (2001)”The semantic web”. Scientific American 284(5) 34–43.

Canfora, G., Penta, M.D., Esposito, R. and Villani, M.L. (2005) “An approach for qos-aware service composition based on genetic algorithms”. In: GECCO., pages 1069–1075.

Cardoso, J. (2007) “Semantic Web Services: Theory, Tools, and Applications”, University of Madeira, Portugal, Information Science reference, édition Dunod, 2007.

Fan, J. B., Ren, and Xiong, L.-R.(2005)” An Approach to Web Service Discovery Based on the Semantics”. In FSKD (2), pages 1103–1106.

Imad, C. (2010) “Using a Genetic Algorithm for the composition of web services”. Master Thesis in computer science. Quebec university at Montreal, 2010.

Jaeger, M. C., Rojec-Goldmann, G., Mühl, G., Liebetruth, C. and Geihs, K. (2005) “Ranked Mat-ching for Service Descriptions using OWL-S”, pages 91–102.

Keller, U., Lara, R., Lausen, H., Polleres, A. and Fensel, D. (2005)" Automatic location of services". In Proc. of the 2nd European Semantic Web Conference (ESWC), pages 1–16, Heraklion, Crete. LNCS 3532, Springer.

Küster, U. and König-Ries, B. (2008) "Evaluating semantic web service matchmaking effectiveness based on graded relevance". In Proc. of the Second International Workshop on Service Matchmaking and Resource Retrieval in the Semantic Web, Karlsruhe, Germany, 2008.

Lecue, F. and Leger, A. (2006) "A formal model for semantic web service composition". In ISWC, the 5th International Semantic Web Conference, November 5–9.

Lecue, F. (2009) "Optimizing OoS-Aware Semantic Web Service Composition". ISWC'09 Proceedings of the 8th International Semantic Web Conference, pages 375–391.

Massimo, P., Takahiro K., Terry R. Payne, and Katia, S.(2002) "Semantic matching of web services capabilities". In, International Semantic Web Conference (ISWC), Sardinia, Italy, 09–12, 333–347.

Srinivasan, N., Paolucci, M. and Sycara K. (2004) "Adding OWL-S to UDDI, implementation and throughput". In the First International Workshop on Semantic Web Services and Web Process Composition (SWSWPC 2004), pages 6–9.

Schumacher, M., Helin, H. and Schuldt. H.(2008) "Semantic Web Service Coordination". Chapter 4, CASCOM: Intelligent Service Coordination in the Semantic Web, Birkhäuser Basel.

Stollberg, M., Keller, U., Lausen, H. and Heymans, S.(2007) "Two-Phase Web Service Discovery Based on Rich Functional Descriptions". In ESWC '07: Proc. of the 4th European conference on The Semantic Web, pages 99–113, Berlin, Heidelberg.

Talantikite Nacer, H., Aissani, D. And Boudjlida, N. (2009) "Semantic annotations for web services discovery and composition," Computer Standards Interfaces, vol. 31, no. 6, pp. 1108–1117.

Thomas, W., Steffen, B. and Kurt, G. (2007) "Web Service Composition Systems for the web Service Challenge – A Detailed Review". University of Kassel, Wilhelmsh¨oher Allee 73, 34121 Kassel, Germany.

PART II

MOBILE TECHNOLOGY

CHAPTER ELEVEN

COMPARATIVE STUDY OF LOAD BALANCING, SELF-OPTIMIZING SCHEMES IN LTE NETWORKS

FATIMA EL MORABET, SANAE CHAIRI, AHMED EL MOUSSAOUI AND NOURA AKNIN

3GPP LTE network is a promising candidate for the next generation of wireless networks. This future mobile communication system will benefit from a significant degree of self-organization. The main drivers for self-organization networks (SON) are the large number of highly complex network parameters, which require a high degree of maintenance in the optimization process. Another challenge is the coexistence of several mobile network technologies that will be operated in parallel (e.g., 2G, 3G and LTE/SAE) (European Commission Information Society and Media 2008, Dottling and Viering 2009). The key objectives for the deployment of SON in wireless access networks are as follows.

- Achieving substantial operational and capital expenditure (O/CAPEX) reductions, a result of less manual intervention in the network operational and optimization tasks. Thus, more users can be accommodated into the system.
- Enhancing network capacity, coverage, service quality and resource utilisation.

The SON mechanism is divided into three domains: self-configuration, self-optimization and self healing. Each of these domains contains some use cases. The use cases are a newly established concept of describing what a solution to a particular problem could attain. Among use cases of self-optimization, the Mobility Load Balancing problem is a significant self optimizing challenge faced by LTE networks, due to the rapid developement of network applications and services. These developments cause resource shortages in cellular networks, time-varying and random traffic loads, and utilize resources efficiently (Hu et al. 2010). The goal of LB process is to cope with undesired effects of load imbalance and to reduce the number of handovers and redirections needed to achieve balance (3GPP [TS 32.522] 2011). The load balancing mechanism aims at shifting some of the traffic load from highly loaded cells to lightly loaded neighbors in order to make use of the radio resource more efficiently across the whole network and increase performance. The rest of the paper is organized as follows.

- Section II gives an overview of SON and LB Mechanisms.
- Section III presents the Load-balancing self-optimization algorithms in LTE system.
- Section IV proposes our LB algorithm.
- Section V presents the main conclusions of the study.

Overview of SON and LB Mechanisms

With the rapid development of wireless communications systems, self-organization networks have become indispensable. This technology is a crucial challenge faced by LTE networks due to the technical and economical incentives. These include the huge number of network elements, the fast growth of network applications, and the reduction of capital and operational costs. Another challenge is the parallel operation of multiple radio networks such as 2G/3G networks and LTE.

However, with these heterogeneous technologies scenarios, it will be difficult to manage and optimize the variety of deployed networks and their specific operational paradigms. The SON concept was widwly used after it was adopted by the Next Generation Mobile Networks (NGMN) alliance to deal with foreseen challenges. The SON mechanism can be divided into three categories: Self-Configuration, Self-Optimization and Self-Healing. These categories contain some cases of use.

SON Use Cases in 3GPP

Self-configuration: The self-configuration function operates in the pre-operational stage of a wireless mobile network. This process enables fast configuration and the deployment of network elements, reducing the capital expenditures and deployment time. The self configuration use cases are defined as follows (3GPP (TR 36.902) 2010, 3GPP (TS 32.500) 2008):

- Self-establishment of a new eNodeB
- Automatic Neighbor Relation (ANR) Management
- Automated configuration of Physical Cell Identity (PCI).

Self-optimization: The self-optimization process works in operational phases. This function allows the mobile network to dynamically adjust and optimize algorithms and network parameters. Also, the self-optimization techniques can minimize the manual optimization tasks, which reduce the operational costs. According to 3GPP (TR 36.902) 2010, 3GPP (TS 32.500) 2008), the most important use cases of self-optimization are:

- Coverage and capacity optimization.
- Energy savings.
- Interference reduction.
- Mobility robustness optimization.
- Mobility load balancing optimization.
- RACH optimization.
- Inter-cell interference coordination.
- Continuous optimization due to dynamic changes in the network.
- Optimization of Quality of Service (QoS) related radio parameters.

Self-healing: The goal of Self-healing technique is to automatically localize and detect most of the failures, and after that, solve these failures. This capability of SON technology will enable mobile network operators to adjust radio parameters and algorithms in neighboring cells. The principal self- healing use cases are defined as follows:

- Self-Recovery of network element software.
- Self-Healing of Board Faults.
- Cell Outage management.

Load Balancing Optimization

From the previous generation of mobile networks the load imbalance remains a critical problem that reduces the performance of a network; previously such problems were solved manually. For example in UMTS network this problem was solved by soft handover functionality. While in the second generation (e.g,. GSM) the load is balanced manually via BSC that adjusts the control parameters for multiple BTSs (SOCRATES (Deliverable D 5.9) 2009).

The load balancing self-optimization use case can adjust automatically the handover parameters and the cell reselection. The principal objective of LB is to cope with uneven distribution of traffic load and local congestion while increasing the overall system capacity and service coverage, as well as reducing operational costs for network management and optimization tasks.

The load balancing is applicable to the following scenarios (R3-072249- 2007):

- Intra-frequency LTE
- Inter-frequency LTE
- Inter-RAT

Load-Balancing Self Optimization Algorithms in an LTE System

The development and implementation of algorithms is a crucial task in load balancing self-optimization, the cause of large number vector of input data, and a huge number of network parameters that complicate the search approaches. Also the decision is difficult due to erroneous information. In addition, it's impossible to do the error test in a live network due to the risk of negative performance effect.

Several researchers have been proposing algorithms to solve the imbalance load problem in an LTE network as well as in a Heterogeneous Network; in the following, we are going to discuss some of these works.

Algorithms Based Handover Parameters Optimization

The algorithm proposed in Lobinger et al. (2010) evaluates the load condition in a cell and the neighboring cells and estimates the impact of changing the HO (handover) parameters in order to improve the overall performance of the network. The proposed method estimates the load after HO would occur, which is based on SINR (signal-to-interference and noise ratio) prediction and UE (User Equipment) measurements. A load balancing algorithm can redistribute the load by dynamic changing of cell-specific handover offset. The simulation results suggest a significant decrease in the average number of satisfied users in function of time when using load-balancing mechanism.

Kwan et al. (2010) proposes an intra-frequency load-balancing algorithm based on automatic adjustment of handover thresholds. This study is concerned with the event A3 (3GPP (TS 36.331) 2009), in which the handover is detected when a neighboring cell offers a better signal quality for a particular UE than its currently serving cell. The results showed significant gains in the call blocking rate and in cell-edge user throughput.

Other studies that redistribute the load through handover parameters optimization are presented in Zhang et al. (2010), Nasri and Altman (2007), Hu et al. (2010), Turkka et al. (2011), and Szilagyi et al. (2012).

Figure 1: Organigram of load balancing algorithm

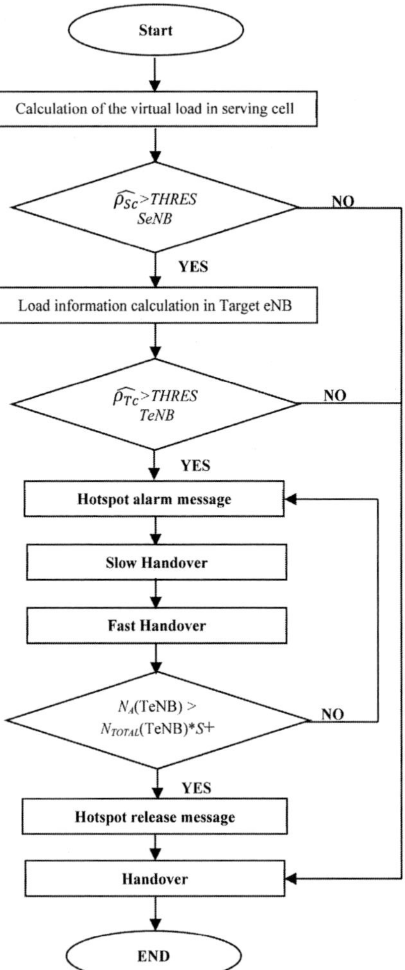

Algorithms Based on Time to Trigger (TTT) Parameter Optimization

The research in Kim et al. (2007) proposes a scheme that balances the load in NGN by dynamically controlling the time point of handover based on the amount of load of the current serving cell and the target cell. The simulation results indicate that the proposed algorithm minimizes the number of overloaded cells and supports higher service quality than conventional handover.

Proposed Load Balancing Algorithm

In this section we are going to explain and discuss our proposition scheme to redistribute the load. The work proposition is based on the dynamic adjustment of cell-specific handover offset as well as time to trigger parameter. The scheme will be based on the SINR of every user, and UE measurements. Based on Viering et al. (2009) the target functions used in this study are expressed as follows:

Signal-to-Interference and Noise Ratio (SINR)

$$SINR = \frac{P_{X(u)} * L_c(\vec{q}_u)}{N + \sum_{c \neq X(u)} \rho_c * P_c * L_c(\vec{q}_u)} \quad (1)$$

Where Px(u) is the transmit power of the serving, eNB, $L_c(\vec{q}_u)$ is the overall path loss, and N is the thermal noise.

Virtual Load

The overload cell appears when the available PRBs (physical resource blocks) are not enough to support the requirements of users. The virtual cell load $\widehat{\rho_c}$ is a newly introduced concept which is defined as the sum of the required resources of all serving users in cell c, divided by Ntot.

$$\widehat{\rho_c} = \frac{1}{N_{PRB}} \sum_{u|X(u)=c} N_u \quad (2)$$

It can be determined that if $\widehat{\rho_c} \leq 1$, all users in the cell are satisfed. However, if $\widehat{\rho_c} \geq 1$, it means only $1/\widehat{\rho_c}$ of the users are satisfied (e.g, $\widehat{\rho_c}$=4 means only 1/4 of the users are satisfied).

Unsatisfied Users Metric

The unsatisfied users metric determines the number of users that cannot achieve the required service quality due to resource limitations or transmission power limitations. The total number of unsatisfied users in the whole network can be written as:

$$Z = \sum_{\forall c} \max(O, Mc^* \left(1 - \frac{1}{\widehat{\rho_c}}\right)) \qquad (3)$$

Where M_c is the total number of users in cell c, and $\widehat{\rho_c}$ the virtual load in cell *c*.

Conclusion

In this paper we have presented an overview of SON and LB Mechanisms. Also, we have proposed some examples from literature that are concerned with load balancing algorithms, and have proposed our algorithm. This algorithm controls the load in the serving cell as well as in the neighboring cell. When the serving cell detects overload traffic, it communicates with a neighbor cell to initiate the handover time scheme. According to the response of neighbor cell, the serving cell determines the handover actions that will be triggered to balance load. While developing this research and obtaining simulation results, we expected that such an algorithm can reduce the blocking, dropping probability as well as enhance the quality of service of an LTE mobile network.

References

3GPP, TR 36.902, 3rd Generation Partnership Project; Technical Specification Group Radio Access Network; Evolved Universal Terrestrial Radio Access Network (E-UTRAN); (2010). "Self-configuring and self-optimizing network (SON) use cases and solutions"; (Release 9) V9.2.0.

3GPP, TS 32.500, 3rd Generation Partnership Project; Technical Specification Group Services and System Aspects; Telecommunication Management; (2008) "Self-Organizing Networks (SON); Concepts and requirements"; (Release 8) V8.0.0.

3GPP, TS 32.522; 3rd Generation Partnership Project; Technical Specification Group Services and System Aspects; Telecommunication management; (2011) "Self-Organizing Networks (SON) Policy

Network Resource Model (NRM) Integration Reference Point (IRP) Information Service (IS)"; (Release 11) V11.0.0;

3GPP, TS 36.331 3rd Generation Partnership Project; Technical SpecificationGroup Radio Access Network; (2009), "Evolved Universal Terrestrial RadioAccess (E-UTRA); Radio Resource Control (RRC); Protocol specification (Release 8) V9.1.0.

Dottling, M. and Viering, I. (2009) "Challenges in mobile network operation: Towards selfoptimizing networks", in Proceedings of IEEE International Conference on Acoustics, Speech and Signal Processing, Taipei, Taiwan, pp 3609–3612.

European Commission Information Society and Media, (2008) "Self-optimisation and self-configuration in wireless networks", European Research Project, available at http://www.fp7 socrates.eu.org.

Hu, H., Zhang, J., Zheng, X., Y. Yang and P. Wu; (2010) "Self Configuration and Self-Optimization for LTE Networks", IEEE Communications Magazine Volume: 48 Issue: 2, pp 94-100.

Kim, D., Lee, S., Lee, H. and Yoon, H. (2007) "An Effective Hotspot Cell Management Scheme Using Adaptive Handover Time in 4G Mobile Networks", ACM, Volume 17 No. 2, pp 139-154

Kwan, R., Arnott, R., Paterson, R., Trivisonno, R. and Kubota, M. (2010) "On Mobility Load Balancing for LTE Systems" VTC. IEEE 72nd, pp: 1-5.

Lobinger, A., Stefanski, S. Jansen, T.and Balan, I. (2010) "Load Balancing in Downlink LTE Self-Optimizing Networks" VTC. IEEE 71nd, pp 1-5

Nasri, R. and Altman, Z. (2007) "Handover Adaptation for Dynamic Load Balancing in 3GPP Long Term Evolution Systems", in Proc. MoMM, pp 145-154.

Next Generation Mobile Networks, (2007) "Use Cases related to Self Organising Network, Overall Description," Version 2.02.

R3-072249; (2007) "Load balancing use case involving cell reselection andhandover parameters self optimization" Telecom Italia, Orange, TMobile, Telefonica, NTT DoCoMo; Jeju (Korea).

SOCRATES Deliverable D 5.9, (2009) "Final Report on Self-Organisation and its Implications in Wireless Access Networks".

Szilagyi, P., Vincze, Z. and Vulkan, C. (2012) "Enhanced Mobility Load Balancing Optimisation in LTE", Personal Indoor and Mobile Radio Communications (PIMRC), 2012 IEEE 23rd International Symposium, pp 997-1003.

Turkka, J., Nihtilä, T. and Viering, I., (2011) "Performance of LTE SON Uplink Load Balancing in Non- Regular Networks", Personal Indoor and Mobile Radio Communications (PIMRC), 2011 IEEE 22nd International Symposium, pp 162-166.

Viering, I., Dottling, M. and Lobinger, A. (2009) "A mathematical perspective of self-optimizing wireless networks", ICC'09 Proceedings of the 2009 IEEE international conference on Communications, pp 4320-4325.

Zhang, H., Qiu,X., Meng, L. and Zhang, X. (2010) "Design of distributed and autonomic load balancing for self organization LTE", VTC. IEEE 72nd, pp: 1-5.

CHAPTER TWELVE

CLOUD AS A MAJOR CATALYST OF CHANGE IN CONTEMPORARY BUSINESS ENVIRONMENT

ATIF FARID MOHAMMAD, HAMID MCHEICK AND EMANUEL S. GRANT

The business environment around us is undergoing continuous change at a very fast pace as Cloud computing is taking over. To cope with the issues of process change, such as adaptation of new or different rules and regulations at the time of an expansion of a business, business processes need to be reengineered. This chapter discusses in detail two fundamental concepts of Business Process Reengineering or BPR: Systems Reengineering and Service-Oriented Architecture, also widely known as SOA. This study provides background of these concepts as well as provides the artifacts researched and discusses work by several researchers and business analysts.

The term business process reengineering or BPR was coined by Michael Hammer and James Champy (Hammer and Champy 1993). The two pioneers define business process reengineering as "the fundamental rethinking and radical redesign of business process to achieve dramatic improvements in critical, contemporary measures of performance, such as cost, quality, service, and speed." However, achieving dramatic improvements through business process reengineering is easier said than done. 60-80% of BPR have been unsuccessful. The high failure rate coupled with huge costs make BPR a very risky operation. (Abdolvand, Albadvi, and Ferdowsi 2007). We will also discuss when organizations need to embark on BPR and when they need not follow that path.

System reengineering can be described as focusing on improving existing or legacy systems for the new need or reducing operating cost. In some articles (Feiler 1993, DeBaud 1994), reengineering is closely related to maintenance, which is defined as the modification of a software product after delivery to correct faults, to improve performance or other attributes, or to adapt the product to a changed environment. By placing reengineering on a continuum with maintenance and new development, the true nature of reengineering becomes apparent. Maintenance entails making corrective, perfective, and adaptive changes to software, while development focuses on implementing new capabilities, adding functionality, or making substantial improvements typically by using new computer resources and new software technologies.

Figure 1: Bank customer service solution (Reprinted)

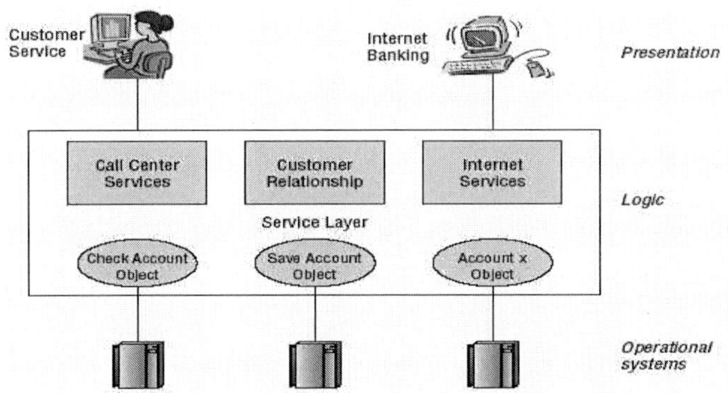

The use of Service-Oriented Architecture or SOA in the Cloud provides services data abstraction. This abstraction can be understood as services messaging metadata, which can be in the shape of XML, XSD or other set industry standards providing interoperability to a user's request. The messaging between services is encapsulated for information-hiding purposes. A SOA solution for a bank as shown in Figure 1 is designed to maintain customer service by providing customer care as well as Web self-serve options for customers.

SOA eases the development of ever-changing applications that compare data with stationary applications while maintaining a decoupled relationship between these applications on a simultaneous basis. Cloud is a loose union of services and SOA has brought in a fresh approach for business information technology departments, making it easy to assemble and configure IT components like building blocks that can be combined to provide easy and fast solutions. For example, a bank provides a line of credit by checking a consumer's credit; an automotive spare-parts seller checks the inventory; Fed-Ex maintains the shipping status for delivery to customers, etc.

This article sheds light on the use of SOA to design the services deployable in the Cloud using business process reengineering to make applications simple and easy to use by both internal and external users of the organizations, which are interested in moving to Cloud computing.

Business Process Reengineering

This section of the chapter presents detailed discussion of the concepts developed by several researchers.

Success Factors of a BPR

Abdolvand et al. (2007) came up with five success factors which they called positive readiness indicators as shown in Figure 2, and they are: egalitarian leadership, collaborative working environment, top management commitment, supportive management, and the use of information technology. The authors based their research on two Iranian companies in different industries as well as research done by other experts. Each main factor has sub-factors. The combination of the factors and their sub-factors cover all the important factors (Abdolvand et al. 2007).

Figure 2: Business Process Factors

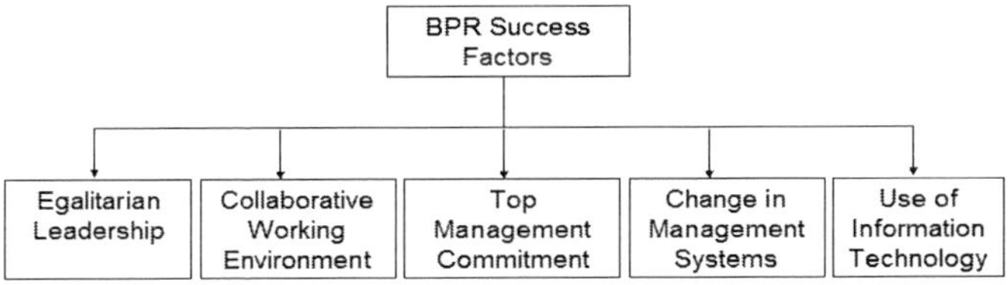

A critical success factor is what Abdolvand et al. (2007) called egalitarian leadership. Top management should drive the changes by providing vision and employees should become more responsive. However, while communication is very important in the success of any venture, culture plays an important role in how people communicate. For example, in low power distance cultures such as the United States and some West European cultures, relations between leaders and the average citizen is much closer than in other cultures that have high power distance. Such cultures (high power distance cultures) always expect ideas to come from the top in the form of commands, and the rest are expected to obey without questioning the wisdom of such ideas. Would this egalitarian leadership approach work universally? We believe it is highly unlikely to work across cultures.

The third and fourth success factors are top management commitment and supportive management. Management should have a clear strategic vision, or, in other words, they must know where the organization is and where it is headed (Abdolvand et al. 2007). The authors were short on description on this topic. However, we believe without managerial commitment and involvement, the organization could easily fail to realize its objectives. Management needs to lead the way the path and offer support to their employees in order to transition seamlessly into their new environment. For example, one way management could support its employees is to provide training.

The fifth success factor is the use of information technology. As per Abdolvand el al. (2007), “IT is introduced as a critical component and even a natural partner of BPR, which has a continuous and important role in BPR projects”. We agree with the authors’ findings. In fact, IT is a natural “ally” of BPR. It is intrinsically an integral part of the process. However, organizations need to make due diligence before undertaking the proposed change. They must calculate their ROI before investing in IT. If you can’t recoup your investments in a reasonable time, don’t invest in the proposed project.

Failure Factors

Abdolvand et al. (2007) cited resistance to change as the most common factor that causes the failure of BPR. The authors called this factor the “unreadiness factor”. Humans resist change, particularly when they don’t expect to benefit from the change. We believe since humans are self-interested, and some employees may even go to great lengths to derail the process, management must reassure its employees that they won’t be worse off when the process gets completed. The fear factor is a legitimate concern which needs to be addressed before the change starts.

Another factor that could contribute to BPR’s failure is having unreasonable expectations about the project. According to Herzog, Polajnar, and Stefano (2007), “The first consists of training about the importance and role of BPR, about the benefits of BPR, the role of cooperation, the commitment to employee training, and the resource availability for training”. Herzog et al. (2007) argue that management must be knowledgeable and have realistic expectations about the benefits of BPR and what it can actually contribute to the bottom-line.

One often overlooked factor in BPR failure is culture. We already mentioned some success factors that may not work across all cultures. According to Newman and Zhao (2008), "The tenets of Confucianism suggest that a BPR effort will challenge Chinese ideology and the workforce will resist radical change including changes to the management hierarchy. Confucius considered it important to maintain a state of harmonious equilibrium in society". The authors found that Chinese companies considered cultural issues as the most important factor when undertaking BPR. Newman and Zhao mention that Chinese culture prefers gradual change, and BPR advocates a dramatic change; hence, most companies would not adopt BPR. Moreover, due to the risk-averse nature of Chinese society, Newman and Zhao (2008) also have found that "most Chinese managers advocate stability and want to maintain power and control. Indeed, many employees are satisfied with getting explicit work instructions and are not willing to initiate actions that are often associated with risks". We agree with the authors. Usually, uncertainty avoidance cultures, which include Chinese culture, try to avoid taking chances. Such cultures do not like ambiguities. Since BPR is fraught with uncertainty, it may not work in those cultures.

Benefit of BPR

Organizations need to conduct thorough research before they decide on implementing BPR. Having metrics in place ensures that an organization knows what it is getting into. However, in order for reliable metrics to be established, management must be knowledgeable (Newman et al. 2008). There is no black and white answer to this question and experts disagree on this. So, some suggest if the organization is doing well, it should not rush into BPR; making changes gradually might be enough. However, prominent experts in the field such as Michael Hammer advocate no less a change than a complete (100%), fundamental, and radical change. Hammer says don't do changes piecemeal (Hammer et al. 1990).

We suggest that BPR may not be the best approach for some organizations. The factors that could influence an organization's decision whether to implement BPR or not, could be seen after evaluating where the organization stands versus the competition.

System Reengineering

During the last four to five decades systems, systems were being created for industries requirements, the process of design decisions and successful implementation of business rules. In order to effectively use current assets of an organization, it is important to develop a systematic strategy for the continued evolution of older systems to meet changing mission, technology and user needs. Moreover, knowledge of the business rules and technical decisions is often embedded in the code. Such knowledge is difficult to recover after many years of operation, evolution, and personnel change. The software portion of a legacy system may have been written 10 or 20 years ago, developed using archaic and ad hoc methodologies, and sometimes need costly maintenance. The result is a legacy system that lacks the ability to evolve and develop for meeting changing demands. Systems reengineering is an esteemed requirement to fill this gap.

The Goals of System Reengineering

The following Figure 3 shows the goals of system reengineering:

Figure 3: An overview of system reengineering

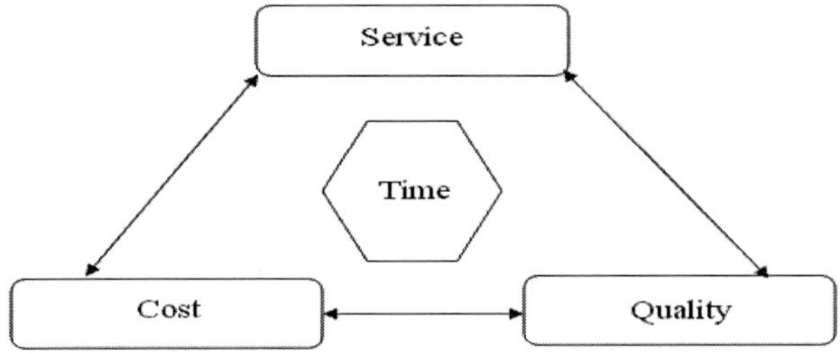

Reengineering is the transformation in a systematic way of an existing system into a new form in order to improve operation by increasing capability, adding new function and raising the level of performance at a lower cost and in less time. The word *system* is a combination of business rules, processes, and operations of user applications. Software is used to elaborate a system in the world of Information Systems.

System Reengineering Approaches

There are three main approaches (Mohammad 2010) utilized to reengineer any system. What follows are details of each. The first one, the "Big Bang" approach, replaces the entire system at one time. The main advantage of this approach is that it can solve an immediate problem, such as migration to different technology, totally different user interface as well as the system architecture. The disadvantage of this approach is that it may not always be suitable for bigger software projects or ERP software such as SAP. For large systems, this approach may consume too many resources or require large amounts of time before the target system is produced. The risk with this approach is high; the system must be functionally intact and work in parallel with the legacy system to assure functionality.

The second approach or "Incremental" approach to a software system re-engineering, is also known as the "Phase-out" approach. In this approach, system fragments are re-engineered and added incrementally as new versions of the system are needed to satisfy new goals that require the training of staff on an individual error-free module one by one. The project is broken into re-engineering fragments based on the existing system's fragments. Every module of the software with a little enhanced functionality on the new technology is developed and tested on the basis of the legacy software module. The main advantage of using this approach is that it has less risk than the "Big Bang" approach as the risks can be identified and monitored at the time of development and beta testing of each module. The disadvantage of using this approach is that it requires a much longer timeframe to reach the goal of reengineered software.

In the "Evolutionary" approach, as in the Incremental approach, fragments of the original system are replaced with newly reengineered system fragments. The components of the legacy system need to be broken down by use and reengineered into new modules. The main advantage is converting of a legacy system to the latest technology. There is one disadvantage and it is that similar functions must be identified throughout the legacy system, then refined as a single functional module.

Using Data Flow to Explain System Reengineering: A Case Study

Figure 4: Legacy system

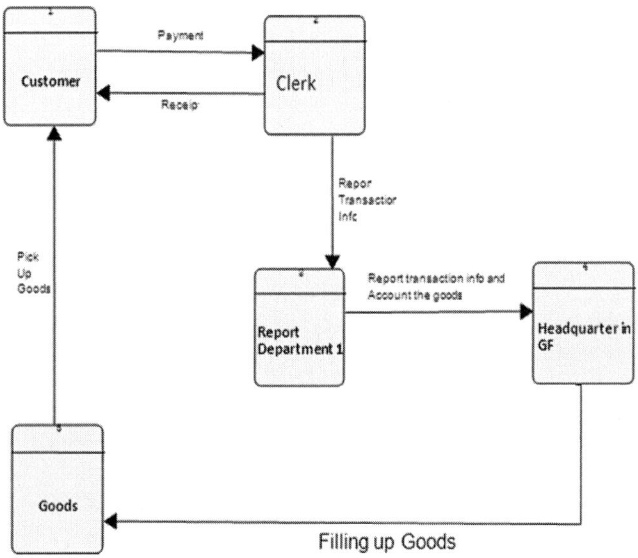

There are quite a few factors involved in the "Migration" process of a legacy system to be converted into a WEB UI system, such as the coding language, base technology that contains RDBMS, operating system, and the new networking technology involving wireless networking as well as advanced hardware infrastructure. System reengineering activities in general require techniques and tools that help to dig out valuable knowledge about the structures and inner mechanism of software systems from the source code using reverse engineering, since other information sources such as system documentation and developer's knowledge are often not available or consistent. A key dilemma during reverse engineering tasks is that the information extracted from the source code has to be condensed and abstracted at a level that can support software engineers' performing essential analyses or modifications of the system and preventing them from an information overflow that makes software reengineering goals complicated. These activities have to be supported by tools. An important consideration in building tool support for reengineering is what information must be provided, and how this information is modeled.

Let us assume that there is a branch of a retail company located in Grand Forks, North Dakota. The data flow of the business process is as follows: First, customer picks up goods in the branch store. After that, customer pays the bill and served by a clerk at front desk. Then, clerk hands in transaction information to another clerk who works in report department and who, after accounting for the goods, reports transaction information to headquarters in Grand Forks. While headquarters received information from branches, it will send goods to the branches as needed. Figure 4 shows the data flow of the business process of this branch of the retail company.

Reengineered System in Retail Company in Grand Forks

After analysis, the system reengineering team translated the legacy system to a Cloud System by using an accounting report system which is sold by an information technology company and operated using a computer. The accounting report system sends transaction information immediately through the website when a transaction occurres. Within AR system Installed, the report department is dismissed automatically, and transaction information would be sent by clerk who works at the front desk the first time. While receiving report transaction information, headquarters is going to restock inventory. This reengineering brought the retail company benefits as described:

a) The service time is reduced, as the clerk would report to headquarters immediately after customer's purchase(s) of the goods for initiating restocking.
b) The cost of goods could be reduced significantly per branch.
c) System reengineering introduced an improved business operations cycle.

Therefore, a basic task of system reengineering is to develop a brief model as a basis for such tools, which can enable the software development team(s) to characterize several associated software applications artifacts needed during the reengineering process and to build high-level abstractions from low level legacy program elements. Problems to be addressed in this project can be as follows:

1. Specification of a model to represent design information of a mini or main-frame system.
2. Specify operations to query the design information stored in the model.

3. Specify operations to abstract low-level program entities into more abstract high-level entities using ERD and DFD mechanisms.

Figure 5: Reengineered Cloud System

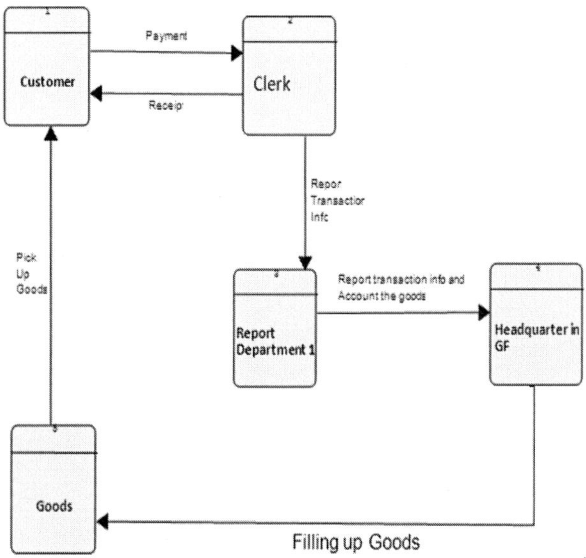

Service Oriented Architecture

SOA provides the framework to work modularly for the Cloud. Due to this flexibility, any business can adapt SOA and assemble its services as needed. A bank's services illustrate this in Figure 1. After changing a customer base or adding new products or services for current customers or to attract new customers, these functionality components can be reassembled. This reconfiguration of service components is invaluable for saving money as well as time. New components can also be added to provide better or more services on the base functionality components. In the case of a merger or cooperative partnership with other related businesses, the components can be included as an additional service to the consumers at one platform. SOA provides the flexibility for businesses to improve and change with the pace of time.

The challenges of the present financial climate can be resolved by software engineers and decision makers, such as CTOs or MIS managers, by aligning business needs, using service components to improve service to customers within a tight budget constraint. SOA is a boon to an enterprise looking to create service components in an agile fashion and reuse an existing components infrastructure. According to Schreiner and Lang (2008), Systems of the future will be based on the concepts of SOA. Service applications will be composed of a number of individual services running at several servers. As illustrated by Erl (2005), service component application logic can be divided into two levels: a service interface, where loosely coupled services are available with their implementation and technology platform; and a service-using application level in which service application logic is developed and deployed on different technology platforms. These services communicate via open protocols.

The purchase planning scenario of a commercial or residential property is an example of a system where the property-selling agency can be provided a flexible approach using SOA. The levels discussed above can be adapted as application and service interface levels. We now take a closer look at how SOA helps to plan a purchase, instantiate, and adapt the composed property purchase service. Initially, three services can be identified as:

(i) "Property for sale" search service
(ii) Offer submission to initiate a purchase service
(iii) Credit check and mortgage service

In this example, partner service implementations are selected as follows. The "property for sale" service can be a local server provided by the real estate agency. The offer submission service can be the Web services provided by another agency, or it can be assumed by law firm service providers. Such legal purchase services may provide the possibility of applying Web service policies, but apart from that, the services might not be configurable. The credit check service is a Web service provided by the financial department of a bank or a mortgage provider. Considering the strict security requirements of financial transactions, these services typically have fixed functional and nonfunctional requirements—i.e., security policies—that cannot be altered.

Due to SOA, these services can be combined at one front-end platform on a website to provide the "property for sale" information and other services as mentioned above. To make this clear, we can take the example of Amazon's store-front. Customers use a browser to get the displays on Amazon.com. The front-end website infers the signals or the customer's intent triggers and calls upon services that do things like acquiring the data for the current on-sale products, or getting the customer's order. The business sector is searching for ways to convert its Information

Technology or Management Information System departments to swiftly adapt to changing business needs. SOA facilitates service orientation.

Challenges of SOA

As the world has become a global village, there is a significant need for businesses to fulfill consumer needs by providing services globally. This pursuit presents some challenges as given below:

- Flexibility is required and there is increasing demand for standardized services with seamless experience for consumers
- Operational cost of these services is reduced by getting satisfactory results due to improved efficiency, which means users control their business, rather than technology
- Services are mostly distributed, getting regulatory approvals
- Rigid information systems wrappers are developed to get the services combined at one framework.
- Efficient use of existing resources is heterogeneous.
- The patterns of services interaction are unpredictable.
- The service(s) "front end" is less useful for testing purposes due to decoupled implementation on the backend.

SOA characterizes and provides the composite framework of the services based on open standards. Multiple services can be combined to serve a consumer at one platform; for example, a travel agent can provide car rental or hotel reservations or can book an entire vacation. This is a convenient way for the consumer to get several services with one phone call or one website utilization. Expedia can be taken here as an example.

Related Work on SOA

SOA provides the methodology to form Web services as functional building blocks to build an accessible infrastructure for consumers which is autonomous and not platform-dependent. Platform independence makes it easier to develop Web services, such as operating systems, programming languages, etc. Quality of Service or QoS is an important factor in SOA as well. We live in an information age, managed by electronic devices. These devices communicate with each other using data communication networks, the best example here being Cloud. Most of the services we use in our daily life are communicated over these networks and are delivered to us on our digital gadgets. Quality of Service (QoS) for networks is an industry-wide set of standards and mechanisms for ensuring high-quality performance for critical applications. By using QoS mechanisms, network administrators can use existing resources efficiently and ensure the required level of service without reactively expanding or over-provisioning their networks. Several researchers (Korostelev et al. 2007, Borcocil et al. (2005), Zeng et al. (2004) have focused on the Web services domain while working on QoSaware service composition and management. During this ongoing research, the composition of QoS-aware services in P2P networks has engrossed much of the researchers' interest. For automatic maintenance of QoS levels research in a P2P infrastructure, Huynh et al. (2007) have proposed self-organized and QoS-aware Virtual Service Communities. Gu and Nahrstedt (2003) examined large-scale P2P systems in detail for their QoS-aware service compositions. They introduced SpiderNet. This framework is capable of doing:

- Service path selection
- Service path instantiation
- Benefit-driven peer clustering

This framework model produced results with a higher success rate of service provision than other problem-solving techniques in which the most appropriate solution of several was found by alternative methods in successive approaches. Let us take an example of an insurance company, which sells the following business services:

Figure 6: A Horizontal Framework of Services

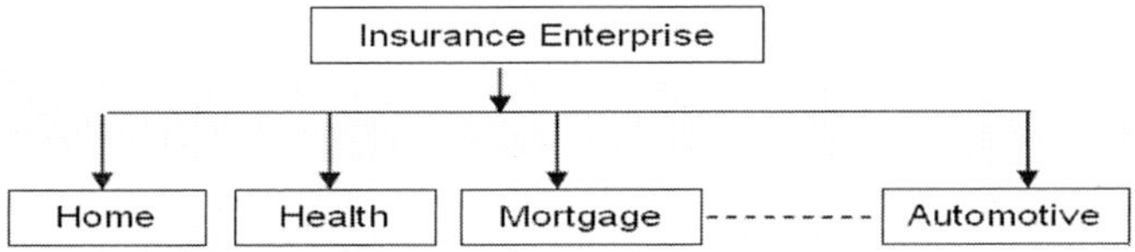

1. Home owner's insurance
2. Health insurance
3. Mortgage insurance
4. Employment insurance
5. Critical illness insurance
6. Automotive insurance

We can see these services as the following framework. Let us explore this horizontal framework of business services, as depicted in Figure 6. All of the mentioned business services are directly related to one enterprise. When a consumer needs one insurance product, the insurance enterprise must evaluate all of the service provisions and locate the appropriate service(s) from the listed services. From the researchers' point of view, a structure for service management is required to maintain QoS in a SOA solution. The aim of this research is to construct a framework that supports QoS for a SOA solution. SOA permits business organizations to interrelate their information systems using predefined rules for mutual dealings within and with consumers (Huynh, Lavinal, and Simoni 2007, Gu and Nahrstedt 2003). SOA is a methodology that interconnects businesses and computational assets, such as software in use, current system applications and human resources of the organizations involved to design, develop, and deploy a SOA solution.

Conclusion

Business process reengineering or BPR is a relatively new business idea. Since its inception in 1990, numerous organizations have tried to adopt BPR. However, the high failure rate indicates that BPR is not beneficial for everyone. Moreover, the cost of failure is very high in terms of both money and morale. Before they try adopting BPR, organizations must take into account where they stand versus the competition. If they are way behind, then they have no choice but to adopt BPR. In addition, since culture plays an important role in determining the success or failure of BPR, organizations that are located in uncertainty avoidance cultures and high power distance cultures must think twice before adopting BPR.

Furthermore, even if all the factors that necessitate the adoption of BPR exist, it would still be hard for most organizations to succeed without a knowledgeable and hands-on management. A knowledgeable management helps set realistic goals and expectations and reduces employee anxiety about the changes to come through effective communication. We would like to conclude that the risks outweigh the benefits, and most organizations must not try to adopt BPR unless they feel they have nothing to lose.

Software architecture is a vital part of software engineering, and the use of Service-Oriented Architectures (SOA) for Cloud deployable services presents an advanced architectural concept with significance for a system's transition from legacy state to a Cloud system state. With SOA-provided flexibility, the new tailorable systems can be produced, and platform independence can also be achieved. Services designed using SOA are formulated software applications, and this formulation is closer to business domains. Research has also indicated that the call for additional metadata of service descriptions is growing quickly, and the amount of data collected from use experiences with a service needs to be stored for analysis of service and its future use. This data handling in terms of storage locations and synchronization raises issues and serious concerns about service performance. The service can have performance issues in terms of message communication netween the user to the service provider due to this additional contextual information.

References

Abdolvand, Neda, Amir Albadvi, and Zahra Ferdowsi. 2007. Assessing Readiness for Business Process Reengineering. Tehran, Iran: Tarbiat Modares University.

Borcocil, E., A. Asgari, N. Butler, T. Ahmed, A. Mehaoua, G. Kourmentzas, and S. Eccles. 2005. "Service Management for End-to-End QoS Multimedia Content Delivery in Heterogeneous Environment." In *Proceeding of Advanced Industrial Conference on Telecommunications/Service Assurance with Partial and Intermittent Resources Conference/E-Learning on Telecommunications Workshop*, 46-52.

DeBaud, J. M. 1994. From Domain Analysis to Object-Oriented Frameworks, A Reuse Oriented Software Engineering Methodology. Atlanta, GA: Coll. of Comput., Georgia Inst. of Technol.

Erl, Thomas. 2005. *Service-Oriented Architecture: Concepts, Technology, and Design*: Prentice Hall.

Feiler, Peter H. 1993. Reengineering: An Engineering Problem. Carnegie Mellon University.

Gu, X., and K. Nahrstedt. 2003. "Chapter 24, QoS-Aware Service Composition for Large-Scale Peer-to-Peer Systems." In *Grid Resource Management: State of the Art and Future Trends*, edited by Jarek Nabrzyski, Jennifer M. Schopf and Jan Weglarz, 395-410. Norwell, MA: Kluwer Academic Press

Hammer, M., and J. Champy. 1993. *Reengineering The Corporation: A Manifesto for Business Revolution*. New York: Harper Business.

Herzog, N. V., A. Polajnar, and T. Stefano. 2007. "Development and Validation of Business Process Reengineering (BPR) Variables: A Survey Research in Slovenian Companies." *International Journal of Production Research* no. 45 (24): 5811-5834.

Huynh, H., E. Lavinal, and N. Simoni. 2007. "A Dynamic QoS Management for Next Generation of Services." In *Third International Conference on Autonomic and Autonomous Systems (ICAS07)*, 11.

Korostelev, A., J. Lukkien, J. Nesvadba, and J. Qian. 2007. "QoS management in distributed service oriented systems." In *Proceedings of the 25th IASTED International Multi-Conference*, 345-352.

Mohammad, Atif F. 2010. "The path from a Legacy System to GUI System." In *Innovations and Advances in Computer Sciences and Engineering*, edited by Tarek Sobh, 141-146. Netherlands: Springer.

Newman, Mike, and Yu Zhao. 2008. "The process of Enterprise Resource Planning Implementation and Business Process Reengineering: Tales from Two Chinese Small and Medium-Sized Enterprises." *Info Systems Journal* no. 18: 405-426.

Schreiner, Rudolf, and Ulrich Lang. 2008. "Protection of complex distributed systems." In *Proceedings of the 2008 workshop on Middleware security*, 7-12. ACM.

Zeng, Liangzhao, B. Benatallah, A.H.H. Ngu, M. Dumas, J. Kalagnanam, and H. Chang. 2004. "QoSAware Middleware for Web Services Composition." *IEEE Transactions On Software Engineering* no. 30 (5): 311-327.

CHAPTER THIRTEEN

SECURITY ATTACKS IN MOBILE AD HOC NETWORKS

AMARA KORBA ABDELAZIZ,
MEHDI NAFAA AND GHANEMI SALIM

Mobile ad hoc networks (MANETs) are becoming an attractive technology for many applications, such as rescue and tactical operations, due to the flexibility provided by their dynamic infrastructure. However, the characteristics and structure of MANETs make them prone to being easily attacked using different techniques often utilized against wired networks as well as new methods specific to MANETs. To design sufficient security solutions for such new environments, we must first understand how MANETs can be attacked. This paper provides a comprehensive survey of security threats in MANETs; we have considered security issues related to both selfish misbehaviors and malicious attacks. In our survey, we investigate selfish misbehaviors and malicious attacks through the different TCP/IP layers, and we discuss how these attacks affect the network. This paper also provides a detailed classification of security issues in MANETs.

Introduction

Mobile Ad hoc Networks (MANETs) are wireless multi-hop networks dynamically constructed by autonomous mobile nodes without the support of any infrastructure or centralized administration. Nodes within each other's radio range communicate directly via wireless links while those which are far apart rely on other nodes to relay messages. This new paradigm of wireless communications aims to make communication possible in some situations where the services offered by both wired networks and WLAN are unavailable. MANETs are mainly useful in military and other tactical applications such as emergency rescues.

Moreover, we can set up an ad hoc network at a conference to distribute files and discuss talks without using any wireless infrastructure that would have to be paid for. Unlike the conventional network, a MANET is characterized by the following features:

- Dynamic Topologies
- Bandwidth-constrained
- Limited Resources
- Decentralized Control

Security Goals

The security services of wireless multi-hop networks are not altogether different from those of other systems. The purpose of these requirements is to protect the exchanged information and nodes' resources from any misuse. The main requirements that effective security architecture must meet are explained in the following:

- Authentication: ensures that the other end of a connection or the originator of a packet is the node that is claimed.
- Confidentiality: ensures certain information is never disclosed to unauthorized entities.
- Integrity: ensures that a packet is not modified during transmission.
- Availability: mainly targets denial of service (DoS) attacks and is the ability to sustain the networking functionalities without any interruption due to security threats.
- Non Repudiation: ensures that the originator of a message cannot deny having sent the message.

Security Attacks

Mobile ad hoc networks are more vulnerable to being attacked than wired networks due to their characteristics. We can distinguish two main categories of attacks: passive and active. A passive attack does not disrupt the operation of the protocol, but attempts to figure out valuable information by listening to traffic. By contrast, an active attack disrupts the operation of the protocol in order to gain unauthorized access, circumscribe availability or degrade the network performance (Cayirci and Rong 2008). Detailed classification of MANET security attacks is shown in Figure 1.

Figure. 1 Classification of security attacks

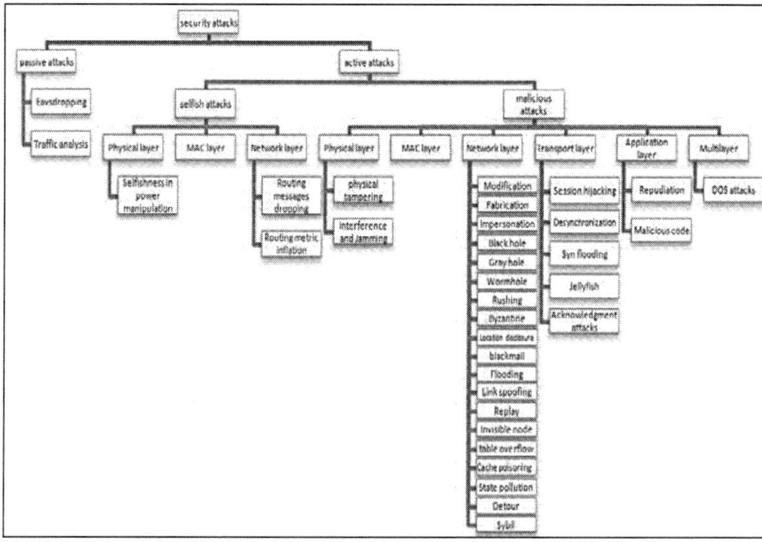

Passive Attacks

Passive attacks do not affect the regular operation of the network; in this case the attacker node is usually invisible to the rest of the nodes and just snoops on the traffic data without altering it. The detection of this attack is very hard since the operation of the network itself is not affected. Examples of passive attacks are eavesdropping and traffic analysis (Pervaiz et al. 2010).

Eavesdropping

Eavesdropping aims to intercept classified data (e.g., credit card number or password) by snooping communication lines. An attacker node can easily intercept data and overhear audio-visual transmissions for the reason that the wireless medium is easy to tap, especially when plaintext, i.e., not encrypted data, is transmitted wirelessly using common standards. For instance, confidential information such as passwords and private keys can be read by the attacker ıf they are transmitted without encryption over the wireless medium (Cayirci and Rong 2008).

Traffic Analysis

The attacker node monitors traffic patterns to infer valuable information such as network topology and sources of certain data packets. The gathered information helps the attacker to figure out the location of events, identities, vulnerabilities, and functions of communication parties. For instance in sensor networks where the nodes closer to the base station relay more packets than the nodes farther from the base station, the attacker analyzing traffic can deduce that the node making more transmissions than the other is closer to the base station and probably is a sink node (Cayirci and Rong 2008).

Active Attacks

In active attacks an attacker node not only listens to the communication but significantly affects the proper function of the network. Active attacks are often detectable except in some cases when the attack is very sophisticated. Active attacks can be classified as selfish or malicious attacks. A selfish attack, also called selfish behavior, targets to gain an unfair share of limited networking resources, even at the expense of other nodes, while a malicious attack intends to disrupt normal network operations (Şen et al. 2010, Wu et al. 2007).

Selfish Attacks

A selfish node does not cooperate in any network functions and exploits the services of the network for its advantage, in order to save its own resources such as battery life. While such misbehavior may not be launched with explicitly bad intentions, it can lead to serious disruptions in network communications such as high route discovery delays and dropped data packets. To render its selfish behavior undetectable, a selfish node can use multiple identities (e.g., Sybil attack) (Alvaro et al. 2004).

Physical Layer Attacks

Selfishness in power manipulation: When a selfish detection system is utilized in the network, such as watchdog (Buchegger et al. 2004), and a selfish node is aware that its behavior is observed, it can hide its selfish behavior and pretend to act cooperatively by transmitting traffic at a power level that is observable to the watchdog and not to other devices (Kim and Jon 2008). As reported in Kim and Jon (2008), if the threshold power to reach the watchdog node is higher than the threshold power to reach the next hop toward destination, the selfish node can transmit with only enough power to reach the watchdog node, instead of transmitting at maximum power, such that the watchdog cannot recognize the selfish behavior.

MAC layer attacks

The IEEE 802.11 CSMA/CA standards (ANSI/IEEE 1999) constitutes a primary MAC protocol for ad hoc networks, which use distributed contention resolution mechanisms to coordinate access to the wireless channel. There are several manipulations violating the rules of the MAC protocol that selfish node can perform to maximize its throughput and conserve its energy. Below is a list of some possible scenarios discussed in Guang et al. (2008) and Kyasanur and Vaidya (2005) that a selfish node can perform:

- It can adjust its backoff time for a constant short period equal to (contention window minimum) CWmin or less at all times.
- It can decrease the size of interframe spaces (both SIFS and DIFS).
- It can selectively scramble MAC frames sent by other nodes in order to obligate them to increase their CW.
- It can delay CTS and ACK, or reject RTS and DATA, so the sender doubles its CW and consequently it gets to transmit.
- It may increase its network allocation vector (NAV) value of the duration field of authentic RTS, CTS frames, in order to prevent its neighbors from competing with it during this period.
- It may also transmit when it senses the channel idle before waiting DIFS time, and not double its CW after collision.

Network layer attacks

Due to the infrastructureless nature of ad hoc networks, where each node is supposed to relay other node's packets, the selfish node has more opportunities to escape performing any function requested by the other nodes.

Routing messages dropping: The selfish node intentionally drops routing packets that are not destined for it, or forwards those packets but with a time-to-live (TTL) of 0 to prevent the creation of routes. Thereby the selfish node can avoid forwarding many subsequent data packets (Kim and Jon 2008). This attack may reduce network performance and prevent end-to-end communications between nodes, if the dropping node is at a critical point in the network (Şen et al. 2010).

Routing metric inflation: The selfish node may make routes through itself appear longer than they actually are by increasing hop counts so the source nodes are more likely to select other routes that seem to be shorter. Selfish node can also generate packets that advertise arbitrarily high distance to a given destination, or stop announcing updates that contain better routes that pass through it.

In routing protocols that suppress duplicate packets, such as DSR (Johnson et al. 2001) and AODV (Perkins et al. 2003), each node forwards only the first received packet and deletes any later copies of the same packet. The selfish node can break this rule by waiting to receive several duplicates and then forwarding the packet with the highest routing metric, decreasing the probability of being selected in the discovered route (Kim and Jon 2008).

Malicious Attacks

Physical layer attacks

Physical tampering: An attacker may either tamper with the physical device and extract sensitive information such as cryptographic keys to gain unrestricted access to higher levels of communication, or physically damage hardware to exclude certain nodes.

> "This attack can also be considered to fall in the domain of fault tolerance, which is the capability to maintain networking functionalities without any interruption caused by node failures" (Cayirci and Rong 2008).

Interference and Jamming: An attacker transmits spurious signals to interfere with the radio frequencies used by network nodes. If the attacker has a powerful transmitter, a generated signal will be strong enough to overwhelm the targeted signals and disrupt communications (Cayirci and Rong 2008). The most *frequent* forms of this attack are random noise and pulse. The jamming attack can be mounted from a location remote to the target networks, and can put N nodes out of service with K randomly distributed jamming nodes, where K is much less than N (Wu et al. 2007).

Mac layer attacks

A malicious node uses different misbehavior techniques, without large deviation from the standard, in order to cause damage to its neighbors. Below there are some possible scenarios that malicious node can perform (Guang et al. 2008):

- "Whenever an RTS signal is received, a signal that collides with the CTS signal is transmitted. Since the nodes cannot start transmitting data before receiving the CTS, they continue sending RTS signals" (Cayirci and Rong 2008).

- "If the MAC scheme is based on sleeping and active periods, jamming only the active periods can continuously block the channel" (Cayirci and Rong 2008).
- "False RTS or CTS signals with long data transmission parameters are continuously sent out, which makes the other nodes that do virtual carrier sensing wait forever" (Cayirci and Rong 2008).
- Increasing the NAV value of the duration field of authentic RTS and CTS frames will silence all its neighbors for a longer period than the required time for the ongoing transmission (Kim and Jon 2008).
- Colluding malicious nodes can continuously send data to each other in order to exhaust the channel capacity in their proximity and hence prevent other legitimate nodes from communicating (Guang et al. 2008).
- Deauthenticate neighboring nodes (Alvaro et al. 2004).

Network layer attacks

Modification: Modification is the most common attack in which a malicious node modifies the content fields of routing packets that transit through it. A malicious node could modify packets before rebroadcasting them, so that they include less attractive metrics, false addresses, and fake hop count in order to redirect network traffic. This attack can cause severe routing disruptions such as conflicted and suboptimal routes, erroneous routing table, network partition and loss of connectivity (Pervaiz et al. 2010).

Fabrication: This attack refers to the generation of fake routing messages in order to disrupt network operations or to deplete other nodes' resources. Such attacks are difficult to detect (Hu and Sharma 2005).

Impersonation: Also called spoofing attack, it usually constitutes the first step in the majority of attacks. The malicious node hides its real identity and takes the legitimate node's identity; thus it can receive all the messages destined for this node and gain access to the network. This attack can also be used for creating loops in order to isolate a target node from the rest of the network (Wu et al. 2007).

Black hole: This attack exploits the vulnerabilities of routing protocols and it is carried out in two steps. First, the malicious node attracts traffic through itself by advertising better routes to the requested destinations. Afterward, the malicious node drops all the data or control packets passing through it without any forwarding (Wu et al. 2007).

Gray hole: This attack is a refined form of black hole attack, in which a malicious node drops only selected packets and forwards the others, depending on the source or the destination of packets. Another kind of gray hole may behave maliciously for a given period by dropping all packets then switching to normal behavior later. This attack defeats trust-based mechanisms and makes the detection of the malicious node more difficult to achieve (Xiaopeng and Wei 2007).

Wormhole: Also called tunneling attack, it is one of the most sophisticated attacks in MANETs. In this attack, a malicious node captures packets from one location in a network and tunnels them through an out-of-band channel to another malicious node located several hops away, which replays them to its neighboring nodes, as shown in Figure 2. The tunnel between the malicious nodes is actually faster than links between legitimate nodes, so the tunneled packet arrives sooner than packets through other routes. Therefore, the malicious nodes are more likely to be included in the route and exploit this advantage for future attack. Detection of wormhole attack is generally difficult, and requires the use of an unalterable and independent physical metric, such as time delay or geographical location (Hu and Perrig 2006).

Rushing: This attack can be carried out against on-demand routing protocols that use duplicate suppression in their operations. In order to reduce the route discovery overhead, each intermediate node processes only the first received route request packets and rejects any duplicate packets that arrive later. Rushing node exploits this mechanism by quickly disseminating route request packets in order to be included in the discovered routes. Rushing attacks can be performed in many ways: by transmitting at a higher wireless transmission power level, by ignoring delays at MAC or routing layers, by keeping other nodes' transmission queues full or by using a wormhole tunnel (Hu et al. 2003).

Figure 2. Wormhole attack

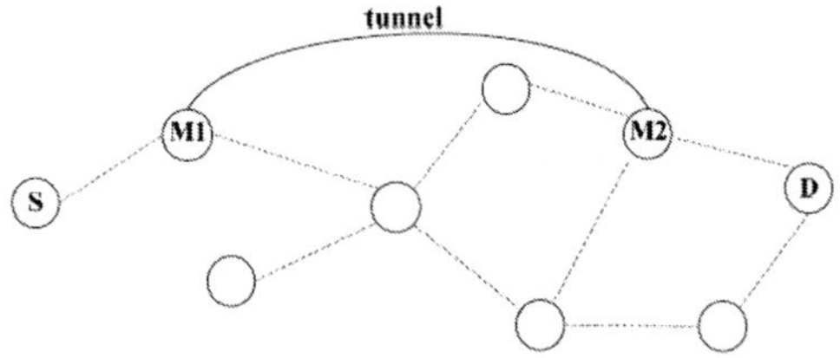

Byzantine: A malicious node or a group of malicious nodes create or modify control packets with false routing information in order to disrupt or degrade the routing operation. This attack is not easy to detect because it does not have a specific form; the malicious node can create routing loops, drop packets selectively or divert packets to non-optimal routes (Wu et al. 2007).

Location disclosure: Location disclosure is an attack that targets the privacy attribute of an ad hoc network. A malicious node can reveal important information such as the location of nodes, or even the structure of the entire network, by the use of traffic analysis techniques, or with simpler probing and monitoring approaches. Colluding nodes may gather information regarding the identities of communication parties, analyze traffic to learn the network traffic pattern, track changes in the traffic pattern, and then plan further attack scenarios. The leakage of such information is devastating in security sensitive scenarios (Argyroudis and O'Mahony 2005, Wu et al. 2007).

Blackmail: This attack occurs against routing protocols that employ malicious detection mechanisms like Watchdog and path-rater. A malicious node may exploit these mechanisms to blackmail legitimate nodes in order to incite other legitimate nodes to put those legitimate nodes in their blacklists (Pervaiz et al. 2010).

Flooding: This is also known as sleep deprivation or resource consumption attack; its goal is to exhaust the network resources, such as bandwidth, and to consume nodes' resources, such as computational and battery power, or to disrupt the routing operation to cause severe degradation in network performance. In this attack a malicious node can flood the whole network with a large number of unnecessary or fake routing control messages, which can lead to severe congestion and battery power consumption, and fill storage space. Alternatively, colluding nodes, at the opposite border of a network, can partition the network by sending a high volume of data between them (Yau et al. 2007).

Link spoofing: In this attack, a malicious node announces fictitious links with a non-neighbor's node. The routing protocol OLSR constitutes a typical example for this attack, where malicious node can convince a victim node to select him as an MPR by advertising fake links with the target's two-hop neighbors (Kannhavong et al. 2007).

Replay: In this attack, a malicious node replicates old control messages that have been intercepted earlier. When the other nodes receive these messages, they update their routing tables according to this expired information. This causes an erroneous view of the network topology and facilitates the impersonation attack. Furthermore, it can be utilized to obtain data which was requested by replayed packet or to disrupt the normal routing operation (Kannhavong et al. 2007).

Invisible node: This attack happens when a malicious node M is situated at the same time in the range of two legitimate nodes N1 and N2, knowing that N1 and N2 are not directly reachable by each other. By relaying the control messages from N1 to N2 and vice versa, the malicious node A arrives to create an nonexistent physical link between N1 and N2, which can fully control and can break at any time it wants. The detection of this attack is very difficult since the malicious node does not alter the relayed messages (Mulert et al. 2012).

Routing table over flow: Basically, this attack targets proactive routing protocols. The malicious node forges routes to a nonexistent destination and then floods these bogus routes with excessive route advertisements to overflow the neighbor's routing table. The goal of this attack is to have enough routes so that the creation of new routes is prevented or the implementation of routing protocol is overwhelmed (Argyroudis and O'Mahony 2005, Hu and Sharma 2005).

Routing table poisoning: Due to promiscuous mode, in which a node can update its routing table with the routing information contained in the packets that it overhears, a malicious node may poison routes to a victim node(s), by generating and sending fabricated routing packets, or by modifying legitimate packets from other nodes, in order to create erroneous entries in the routing tables of the participating nodes. This attack can cause the creation of routing loops, or partition networks or the selection of non-optimal routes (Argyroudis and O'Mahony 2005).

State pollution: In ad hoc networks, each mobile node must be with a free IP address before it may participate in unicast communications. A malicious node can reverse the uniqueness of address assignment by providing incorrect responses regarding the requested parameters, for instance, in best-effort allocations; a malicious allocator can always give the new node an occupied address, leading to repeated broadcast of duplication address detection messages throughout the network and rejection of a new node (Zhou 2008).

Detour: This attack is also called gratuitous detour; it occurs at the discovery phase where the malicious node deflects traffic through suboptimal routes (congested or energy-depleted routes), by altering a route request such that it includes less attractive metrics and becomes more costly beside another route to which the attacker attempts to detour traffic. Different ways can be used for this, like increasing the hop count or delaying rebroadcasting route requests (Yau et al. 2007).

Sybil: In the Sybil attack, a malicious node presents multiple addresses and behaves as if it were a group of nodes. There are, mainly, two different ways through which a Sybil node can get an identity: stealing another node's identity or fabricating fake identities. By impersonating a large number of nodes in the network, the malicious node prevents other nodes from using those addresses, can escape from detection systems, and can "out vote" the well-behaved nodes in collaborative tasks such as Byzantine failure defenses. This attack can severely damage geographic routing protocols, and can even menace multiple path routing schemes and node localization (Cayirci and Rong 2008).

Transport layer attacks

Session hijacking: In this attack, the malicious node usurps the victim's address, deduces the current sequence number used by the target node, and then exhausts the victim's resources to exclude the victim node from the network. Therefore the malicious node can pursue the ongoing session with the target node. Subsequently the malicious node can send injected session data to provoke sequence number conflict, causing the exchange of a large number of ACK packets called TCP ACK storm (Wu et al. 2007).

Desynchronization: The malicious node sends fabricated messages to one end or both ends of communication; those fabricated messages request transmission of missed messages. Thus, the requested messages are retransmitted and if the malicious node sustains an adequate transmission rhythm, it can avoid any useful communication between the two ends. This attack can cause a significant energy consumption of the legitimate nodes in the network (Wood and Stankovic 2002).

Syn flooding: The malicious node attempts to open a large number of new TCP connections with a target node, but without completely establishing the connections, in order to overflow the victim's buffer, such that the victim node cannot accept any other legitimate connection request. The malicious node can maintain these half-open connections by repeatedly sending request connection packets faster than the expiration of pending connections (Wu et al. 2007).

Jellyfish: Jellyfish can be launched against protocols with congestion control techniques, such as TCP and can be performed in three ways. The first one is feasible only if cumulative acknowledgment is utilized, where the malicious node reorders packets during transmission, causing constant retransmissions. The second one is periodic dropping near the Retransmission Time Out (RTO) time-scale, compelling TCP to constantly enter the timeout state. The third one is delaying data packets unnecessarily for random waiting time before forwarding them, causing collisions, significant high end-to-end delay, and packet loss (Mulert et al. 2012).

Acknowledgment attacks: Malicious nodes can jam the segments that convey acknowledgements to terminate connection. It can also forge a fake acknowledgment or increase receiver windows, to cause congestion and degrade the network performance. In some protocols, such as TCP-Reno, which consider the acknowledgment of the same segment multiple times as a negative acknowledgment, malicious node can replay an acknowledgement multiple times to claim a failed delivery (Cayirci and Rong 2008).

Application layer attacks

Application layer is also susceptible to security threats. There exist many attacks applicable at this layer; the most popular are listed below:

Repudiation attack: Repudiation refers to a denial of being involved in all or part of the communication process. For example, the classic repudiation attack is on e-commerce system, in which a dishonest client negates making an online bank transaction or a credit card purchase (Wu et al. 2007).

Malicious code: An attacker can use similar malicious programs utilized in a wired network, as Viruses, spyware, worms, Trojan horses and malicious codes, to achieve its bad intentions. These programs are commonly auto-disseminated throughout network, and can cause severe performance degradation and can even damage both computer system and networks (Wu et al. 2007).

Multilayer attacks

Denial of service: DOS is a multilayer attack; many of the active attacks presented until now may be considered as DOS attacks. According to Cayirci and Rong (2008), DOS is defined as any action that decreases the capacities of networks and prevents it from achieving its intended task properly or in a timely manner. DOS targets the availability of network services, where a malicious node degrades the quality of service of the network by exerting much less effort beside the significant impact made on the network. DDOS is an enhanced version of DOS attack, in which a group of malicious nodes collude to cripple the whole network by flooding it with massive network traffic and therefore consuming all the network resources (Devi and Kannammal 2012).

Conclusion

In this paper, all types of security threats to which MANETs are vulnerable are being presented. A taxonomy of security issues encompassing selfish and malicious misbehaviors, and illustrating to which layer they belong is proposed. MANETs are still in the early stage, and security is such an important feature that it could determine their success and their wide deployment. MANET security constitutes a complex and challenging area, in which research is still being performed and will result in the discovery of new threats as well as the development of new countermeasures.

References

Alvaro, A.C., Radosavac, S. and John, S.B. (2004) "Detection and prevention of MAC layer misbehavior in ad hoc networks", in Proceedings of the 2nd ACM workshop on Security of ad hoc and sensor networks (SASN '04), pp. 17-22.

ANSI/IEEE std 802.11-1999, IEEE 802.11 Wireless LAN media access control (mac) and physical layer (phy) specifications.

Argyroudis, P.G. and O'Mahony, D. (2005) "Secure routing for mobile ad hoc networks", Communications Surveys & Tutorials, IEEE, vol.7, no.3, pp. 2- 21.

Buchegger, S., Tissieres, C. and Boudec, J-Y.L. (2004) "A Test-Bed for Misbehavior Detection in Mobile Ad-hoc Networks How Much Can Watchdogs Really Do?", in Proceedings of the Sixth IEEE Workshop on Mobile Computing Systems and Applications (WMCSA '04), pp. 102-111.

Cayirci, E. and Rong, C. (2008) Security in Wireless Ad Hoc and Sensor Networks. John Wiley & Sons. The Atrium, Southern Gate, Chichester, West Sussex, PO19 8SQ, UK.

Devi, P. and Kannammal, A. (2012) "A hybrid defense mechanism for DDoS attacks using cluster analysis in MANET", in Proceedings of the International Conference on Advances in Computing, Communications and Informatics (ICACCI '12), pp. 287-291.

Guang, L., Assi, C. and Benslimane, A. (2008) "MAC layer misbehavior in wireless networks: challenges and solutions", Wireless Communications, vol.15, no.4, pp.6-14.

Hu, F. and Sharma, N. K. (2005) "Security Considerations in Ad Hoc Sensor Networks", Ad Hoc Networks, 3 (1), pp. 69-89.

Hu, Y.C. and Perrig, A. (2006) "Wormhole Attacks in Wireless Networks", IEEE Journal on Selected Areas in Communications 24 (2), pp. 370-379.

Hu, Y.C., Perrig, A. and Johnson, D. B. (2003) "Rushing attacks and defense in wireless ad hoc network routing protocols", in Proceedings of the 2nd ACM workshop on Wireless security (WiSe '03), pp. 30-40.

Johnson, D.B., Maltz, D.A. and Broch, J. (2001) "DSR: the dynamic source routing protocol for multihop wireless ad hoc networks", in Ad hoc networking, Charles E. Perkins, Ed. Boston, MA: Addison-Wesley Longman Publishing Co., Inc, pp. 139-172.

Kannhavong, B., Nakayama, H., Nemoto, Y., Kato, N. and Jamalipour, A. (2007) "A survey of routing attacks in mobile ad hoc networks", Wireless Communications, vol.14, no.5, pp. 85-91.

Kim, H.J. and Jon, M.P. (2008) "Detecting Selfish Behavior in a Cooperative Commons", in Proceedings of the third IEEE Symposium on New Frontiers in Dynamic Spectrum Access Networks DySPAN, Chicago, Illinois, pp. 1-12.

Kyasanur, P. and Vaidya, N.H. (2005) "Selfish MAC Layer Misbehavior in Wireless Networks," Mobile Computing, IEEE Transactions on, vol.4, no.5, pp. 502- 516.

Mulert, J.V., Welch, I. and Seah, K. G. W. (2012) "Review: Security threats and solutions in MANETs: A case study using AODV and SAODV", Journal of Network and Computer Applications, vol. 35, issue 4, pp. 1249-1259.

Perkins, C. E., Royer, E. M. and Das, S., "Ad hoc On-demand Distance Vector (AODV)", RFC 3561, July 2003.

Pervaiz, M.O., Cardei, M. and Wu, J. (2010) "Routing Security in Ad Hoc Wireless Networks," in Network Security, S. Huang, D. MacCallumand D.Z. Du, Eds. NY: Springer US, pp. 117-142.

Şen, S., Clark, J.A., Tapiador, J.E. (2010) "Security Threats in Mobile Ad Hoc Networks," in Security of Self-Organizing Networks: MANET, WSN, WMN, VANET, Al-Sakib Khan Pathan, Ed. Boca Raton, Florida: Taylor & Francis, pp. 127-146.

Wood, A.D. and Stankovic, J. A. (2002) "Denial of service in sensor Networks", Computer, vol.35, no.10, pp. 54- 62.

Wu, B., Chen, J., Wu, J. and Cardei, M. (2007) "A Survey of Attacks and Countermeasures in Mobile Ad Hoc Networks", in Wireless Network Security, Y. Xiao, X. Shen and D.-Z. Du, Eds. NY: Springer US, pp.103-135.

Xiaopeng, G. and Wei, C. (2007) "A Novel Gray Hole Attack Detection Scheme for Mobile Ad-Hoc Networks", in Proceedings of the 2007 IFIP International Conference on Network and Parallel Computing Workshops (NPC '07), 2007, pp. 209-214.

Yau, P-W., Hu, S. and Mitchell, C. J. (2007) "Malicious attacks on ad hoc network routing protocols", International Journal of Computer Research, vol. 15, no 1, pp. 73-100.

Zhou, H. (2008) "Secure Prophet Address Allocation for Mobile Ad Hoc Networks", in Proceedings of the 2008 IFIP International Conference on Network and Parallel Computing (NPC '08), pp. 60-67.

CHAPTER FOURTEEN

A SMARTER WAY TO IMPLEMENT CIRCUIT SWITCH FALLBACK AND ITS MODIFIED CALL FLOW TO REDUCE THE CALL SETUP TIME

YOUNESS JOUIHRI AND ZOUHAIR GUENNOUN

After the successful testing of the first voice call handover from a 4G long term evolution (LTE) network to a wideband code division multiple access (WCDMA) network using Single Radio Voice Call Continuity (SRVCC) feature (Asmar et al. 2012), the availability of a viable handset supporting this feature in the near future became a certainty. This declaration increases the challenges to operators who still have a voiceless LTE network but are willing to join the group of operators who already have publicized the accessibility of voice services on their LTE networks. The SRVCC solution is the next logical step in the 4G LTE voice roadmap (Asmar et al. 2012), but operators willing to opt directly for this option cannot guarantee their users the availability of voice services, as they do not have any real guaranties about the timing of viable handset availability (Gompa 2012). Also, investing in the deployment of circuit-switched fallback (CSFB) solutions should add significant increases to the cost reserved for voice-over LTE projects, and it might take more time than expected to have handset-supporting SRVCC, especially when adopting the actual way of CSFB implementation.

The remaining sections are outlined as follows. After an overview on CSFB solution in Section I, Section II will explain the CSFB advantages and disadvantages for operators willing to add voice on their LTE networks. Section III studies the existing CSFB implementation and call scenarios having the most noticeable impact on the user experience. Section IV introduces the smart way of deploying CSFB procedures and the modified call flow to overcome the call setup time delay on the most impacted cases. Section V presents our experiment results. Finally, a summary is provided in Section VI.

CSFB Overview: Advantages and Disadvantages for Operators Willing to add Voice on their LTE Networks

The CSFB is one of the interim solutions defined by 3rd Generation Partnership Project (3GPP) to enable voice and other CS-domain services (CS UDI video, LCS, USSD) by reusing the circuit-switched (CS) infrastructure (3GPP 2012), LTE will provide only data services, and for the initiation or the reception of a voice call, fallback to CS domain must be performed (Qunhui 2011). This function is available only in the event LTE network coverage is overlapped by CS network coverage (3GPP 2012). The CSFB functionality provides a seamless voice service without the implementation of voice-over IP (VoIP) in a LTE network (Nokia Siemens Networks 2012).

Unfortunately CSFB suffers from numerous shortcomings and its worst effect is the extra time needed for call setup while receiving or making a call (Kineto Wireless 2009). The call setup is a major factor in determining the quality of experience with phone calls (Kineto Wireless 2009).

The solutions based on SRVCC (One Voice or Voice over LTE via Generic Access) were more suited to playing the alternative role until full LTE coverage is established, and in cases where the VoIP will be the provider of the primary voice services (Jouihri 2012), but the delay on SRVCC implementation on the handset level put a big challenge in front of operators willing to add more services to an LTE network that has already been deployed (Gompa 2012). Adding vital services might inspire many customers who still are confused to opt for this new technology, and it also provides additional advantages that operators can use on their media campaigns to show that they have the best LTE network, especially if we know that these services can generate huge revenue from voice, whch already is proven to be the main revenue-generating service on the legacy network (Jouihri 2012).

After the successful completion, resulting from Qualcomm working with Ericsson, of the first voice call handover from an LTE mobile network to a Wideband Code Division Multiple Access (WCDMA) network using Single Radio Voice Call Continuity (SRVCC) (Asmar et al. 2012), operators are one step closer to getting mobile phone support for SRVCC. The LTE phones supporting SRVCC and using the Snapdragon

S4 MSM8960 multi-mode modem chip likely will arrive either at the end of the first half of this year or at the beginning of the second half of 2012 (Gompa 2012), but there is no official confirmation yet. Also complete testing is required of all the handover cases (from LTE to 2G network and back, from LTE to 3G Universal Mobile Telecommunications System (UMTS) networks and back, from Code Division Multiple Access (CDMA) 2000 networks and back), which might require more time. Even more, operators having multi-technologies must complete the testing of all the functionalities, even the old, with the new mobiles, consult with handset makers if there are any issues detected during the testing phase, and wait for new releases that incorporate the needed corrections before the commercial launch of voice-over LTE services.

Based on the above points, operators willing to add voice functionality to their voiceless LTE networks need to either wait for handset-supporting SRVCC and give their competitors the chance to have precession in the media war, as many operators already have developed voice services support on their LTE network by using alternative solutions such as CSFB for UMTS operators or active dual-mode operation for CDMA2000 operators (Gompa 2012), or these operators might have opted for other alternatives that are not really elegant, such as CSFB (Gompa 2012), which will affect badly the quality and user experience, and also add more cost to any of their projects that involve voice-facility deploying on LTE. Even so, all operators eventually will have to opt for SRVCC solutions as the next logical step in the 4G LTE voice roadmap (Asmar et al. 2012). Actual CSFB implementation also needs time in order to complete the upgrades on all MSC servers, which will make the decision more difficult, as handset makers expect to have the needed phone in a couple of months.

To make choice easier for operators, we first will propose that the CSFB implementation be made in a smarter and simpler manner so that the costs and the time-consumption will be minimal. Then we will take up the biggest problem of CSFB solutions related to the long time needed to establish the calls, check to see if our proposition to simplify CSFB deployment might increase this delay, and study the most-impacted call scenarios with their corresponding call flow. Finally, we will suggest an optimal call flow for the most-impacted call scenario, including our first proposal, in order to reduce the call setup time and provide a comfortable user experience.

So what is the smarter way of deploying a CSFB solution, and which call scenarios have to be improved?

Existing CSFB Implementation and Call Scenarios Having the Most Noticeable Impact on the User Experience

After the LTE network deployment, the legacy 2G/3G network and the LTE network have to co-exist and create a mixed network with many operators (Qualcomm 2012). The mobility management entity (MME) will serve the users in LTE coverage, while in 2G/3G the SGSN will support the data services and the mobile switching center (MSC) server for voice services (Qualcomm 2012). To add the CSFB in a mixed network, the MME should communicate with the MSC server, for the SGs interface was defined to connect the MME and MSC server, and it enables the LTE user's to use a CS network (Qualcomm 2012).

As shown in Figure 1, when we have a mobile call terminating with an LTE user, the MSC server will send a paging request to MME (message 1), and the MME will page the UE (messages 2 &3) (Qualcomm 2012). This page initiates the CSFB procedure, as the user will request the 2G/3G network for transition (message 4). Once transitioned, the normal CS mobile termination call procedure will be performed (Qualcomm 2012).

For mobile-originating calls, the same process will be followed except that the paging procedure will not be needed. When the voice call ends, the user equipment (UE) has to return to LTE network (Qualcomm 2012). Currently, for operators using MSC server and wishing to add CSFB functionality to their networks, they need to upgrade all of their MSC servers so that they support the SGs interface. This interface then allow the legacy network and LTE network to communicate (3GPP 2012).

After the acquisition of voice services, operators have to take into consideration the quality of the user's experience. The additional steps in switching from LTE to 3G for voice calls will impact the call setup time (Qualcomm 2012). It has been concluded that if one selects the correct call- setup method and redirection-based CSFB approaches, and performs detailed measurements in a commercial infrastructure averaged over a variety of good and poor radio conditions, the CSFB call setup increments may not impact the user experience since the measurements fall within the range of 3G call setup times. Even the influence of local signal strength and network conditions can be bigger than CSFB call setup time penalties (Qualcomm 2012). The measurements performed dealt with ideal cases in which the UE will target the same 3G cell that already is stored on the MSC server. To always use the same cell, LTE and 3G access patterns for any given location always must be identical, which can never be the case, as we have different frequencies having different terrain propagation and structure penetration characteristics, as well as potentially different fixed antenna sites (Qualcomm 2012).

Figure 5: CS Fallback Procedure Using SGs Interface In A Mixed Network

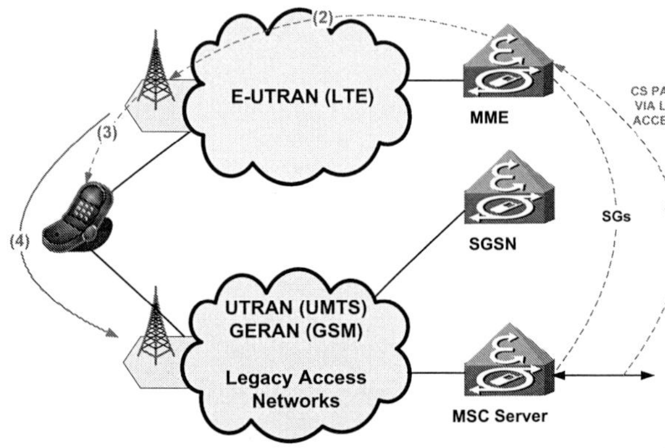

The LTE cells can overlap two or more 3G cells, which can introduce substantial setup delays due to the update of user locations on the MSC server (Qualcomm 2012). This Location Area Update (LAU) procedure might add some seconds delay on the setup time, depending on the load on the network (Qualcomm 2012). The impact will be more if the transition from LTE to 3G happens in an MSC server's "border" area, in which the MSC servers will be involved (Qualcomm 2012).

In this situation, an LAU procedure involving both MSC server (new and old) and HLR should occur to update the UE location information. In this last case, if the CS network does not support the Mobile Terminated Roaming Retry (MTRR) procedure, the call will be cleared (Nokia Siemens Networks 2012). The MTRR procedure might add more than a four-second delay to setup time which will noticeably affect the user experience (Qualcomm 2012).

The current proposal to address the setup delays due to MTRR procedure would involve the deployment of MSC Pool Architecture (Qualcomm 2012), which might result in exceeding the time and the funds reserved for CSFB implementation.

We have concluded that actual CSFB deployment requires the addition of SGs interface support in all MSC servers, and also that the user experience will be impacted most by calls using the Mobile Terminated Roaming Retry procedure.

So how can we make the CSFB implementation simple? And which modifications can we suggest to improve the call setup time of MTRR procedure?

The Smart Way of Deploying CSFB Procedure and the Modified Call Flow to Overcome the Call Setup Time Delay on the Most Impacted Cases

The Easy and Smart Way of Deploying CSFB Procedure

For operators having a mixed (LTE & 3G/2G) network and using MSC servers, CSFB deployment requires a software upgrade in all MSC servers to enable them to communicate with the LTE network (MME) by adding the support of SGs interface (3GPP 2012).

The upgrade of the entire network may drain all the resources reserved for a voice project on an LTE network. This upgrade will provide just a temporary solution for voice services (Asmar et al. 2012), as it bypasses the still-needed SRVCC solution (Gompa 2012) into which operators will have to invest the majority of their voice project budget. The entire CSFB deployment requires time, and the cost will be higher depending on the number of MSC servers on the network. The implementation likewise will cause downtime on live services, which might result in lost revenue and provide a chance for competitive operators to catch roamers moved from MSC servers towards their network during the downtime. The effect will be more visible for operators having big networks and huge numbers of MSC servers.

To make the CSFB implementation easy and simple, we should find a way to compensate for the entire implementation and replace the larger general change with a limited one. The idea of having proxy equipment to enable the needed functionalities seems to be self-evident, but the smart selection of this equipment might save time and reduce the cost. Normally, all the MSC servers are connected to each other using BICC or SIP connections, so if we implement the CSFB procedure on one of them and enhance this implementation to support the traffic coming from all other MSC servers by increasing the signaling capacity between MME and the MSC server having a CSFB solution, we will save the cost of adding new equipment, and we can save the time by using the existing connection to provide CSFB services to other MSC servers. The MSC server having this functionality will play the role of proxy for CSFB services and will be called proxy MSS; we can have more than one proxy MSS (as shown in Figure 2), depending on the traffic load and operator decisions, but as a first step, we can start with one MSC server. We also have to be smart when selecting the MSC server because the addition of proxy MSS will add additional time to the CSFB call setup, as the call has to be sent to proxy MSS, and from there, the paging request will be sent to MME. In the case of full CSFB implementation, the paging can be sent directly from the MSC server initiating the call to the MME (3GPP 2012). The selected MSC server should be the one usually serving the biggest number of subscribers and having the largest number of connections to the radio network. In this way, we can reduce the number of subscribers impacted by this additional time due to proxy MSS, and we also will thus minimize the cases where location update will happen in different MSC servers during the CSFB procedure and reduce the need for the MTRR procedure.

The proxy MSS seems to be the best option for enabling the CSFB solution without impacting significantly the existing 3G/2G network but, as discussed before, the user's experience will be badly affected on the MSC Server "border" area, where the MTRR procedure might add more than 4 second delay to the call setup (Qualcomm 2012). The use case that we will try to enhance is the call scenario in which the MTRR procedure is needed and in which the paging request will be forwarded from the proxy MSS to MME. By proposing some modifications to the general MTRR call flow, many messages can be removed. In this way we can compensate for the extra time added by the proxy MSS addition and, even more importantly, we can reduce the actual long call setup time needed to perform the MTRR procedure that is badly affecting the users' quality experience (Qualcomm 2012), without requiring any expensive solutions, such as pooling architecture. These modifications will be based on the addition of extra information to the normal signaling messages, so the entity receiving these messages will get the complete information and will not wait for a new message to perform the next task.

Figure 2: Proxy MSS Architecture

The Modified Call Flow of the Most Impacted Use Case Including the Addition of Proxy MSS

Our optimized call flow will be based on the use case in which the UE camps, during the CFSB procedure, on a GSM/Edge Radio Access Network on a GSM/Edge Radio Access Network (GERAN)/Universal Terrestrial Radio Access Network (UTRAN) cell served by a different MSS than the one having the user device registration. In the current solution, a location update procedure should occur. Also, the Mobile Terminated Roaming Retry procedure is needed to terminate the call successfully (Nokia Siemens Networks 2012).

As shown in Figure 3 for the normal MTRR scenario, if a Mobile Terminating call arrives to a GMSC, the GMSC will perform an HLR inquiry by sending a Send Routing Information (SRI) to the home location register (HLR).

The HLR will send a Provide Roaming Number (PRN) message to get the Mobile Subscriber Roaming Number (MSRN) from the MSS, the MSS will provide a MSRN on the PRN response, and the HLR will forward the MSRN to the GMSC on the SRI response. The GMSC will initiate an Initial Address Message (IAM) based on the received MSRN towards the MSC server. Once received, the MSC server will send a paging over SGs request to the MME, and the MME will page the UE. Our recommendation here is to add the paging request on the PRN message sent from the HLR to the proxy MSS, as shown in Figure 4. Once the HLR finds that the subscriber is roaming in the LTE network, the proxy MSS will send the paging directly over SGs request to the MME. The received IAM message from GMSC to trigger the paging procedure will not be required (Figure 3).

The paging over the SGs interface would initiate a possibly long procedure of CS Fallback, and so we must either increase the PRN timer on the HLR or make a separate timer for the PRN request so that our call will not be cleared due to PRN timer expiry.

After that, for both cases the MME will initiate a paging procedure by sending a SERVICE-REQUEST message to the Mobile Switching Center (MSS)/Visitor Location Register (VLR) or proxy MSS, which will be taken as an intermediate paging response and will be used to stop the paging timer (3GPP 2012). A new timer will start to supervise the final paging response from the terminating subscriber, at this point; the subscriber may have the option of deciding whether to accept the call or not (3GPP 2012). If the subscriber accepts the call, following our use case, the UE camps on a GERAN/UTRAN cell served by a different MSS than the one having it is registration.

The UE executes fallback to GERAN/UTRAN and sends a Location Update Request to the new visit MSS (VMSS), which controls the target GERAN/UTRAN access. The new VMSS initiates an update location procedure to the HLR and the HLR sends a MAP Cancel Location to the old MSS in normal MTRR case or to the proxy MSS in our use case (Figure 3 & Figure 4). In our use case, once the proxy MSS receives the Cancel Location message, it recognizes that the UE executed a fallback to another VMSS. The proxy MSS stops the paging over the SGs interfaces and sends an SRI message to the HLR to obtain routing information (MSRN) from the new VMSS. There will be no need for RCH and a release message, as these messages are used to release the connection between GMSC and the MSS initiating the paging procedure, after receiving this request in an IAM message from GMSC.

But in our use case, the proxy MSS starts the paging directly after receiving the PRN including this information, which eliminates the need for the GMSC IAM message to initiate the paging (as described before), so there is no open connection between GMSC and the proxy MSS; therefore, it will not require any RCH and release messages to close it.

Next, for both cases, once the HLR receives the SRI, either from GMSC or the proxy MSS, it sends the PRN to the new VMSS. The new VMSS answers with an MSRN in the response. The proxy MSS or GMSC initiates an IAM message based on the received MSRN.

In the normal MTRR case, after receiving the IAM message, the new MSS will start a paging over the A/Iu interface, and the UE will reply back with a paging response. From this point, the call establishment procedure is identical to a standard CS Mobile Terminating Call (Figure 3).

In our use case, we have recommended using the flag indicating the pending state of any mobile terminating call on the new MSS. So during the location update request initiated by the UE, as a result of CS fallback to the GERAN/UTRAN procedure, MSS must check for any pending terminating CS calls and, if the flag is set, maintain the radio connection after the Location Update procedure has been completed for the pending call. The MSS might start a timer to supervise the arrival of call setup, meaning that, after receiving the IAM message coming from the proxy MSS to the new MSS, the MSS stops the timer and establishes the CS Mobile Terminating Call as usual, reusing the signalling connection that has been maintained. In this scenario there will be no need for new paging over the A/Iu interface. As we know, the paging procedure is a time-consuming operation and in many cases, the re-paging action is needed to reach the UE which will double the operation time. The call setup time can be reduced considerably by adopting this recommendation.

Experiment Results

The measurements of call setup time have been collected from live 3G networks with commercial infrastructure. Our recommendation, to modify the CSFB call flow cannot be implemented directly and easily on a live network. The values of call setup time after change have been measured by simulating the same previous calls, done on live network while reusing the exact duration taken by each message, until we have composed the complete modified call flow respective to our recommendations. With the proposed call flow, the IAM message used to trigger the paging over SGs will be removed along with its corresponding RCH and release messages. Also, the paging procedure over the Iu interface will not be required.

As mentioned before, the paging procedure is a time-consuming operation. Table 1 shows that the re-paging action was needed in most of the calls performed during the testing, which added a penalty to the call setup time.

From Figure 5, we can conclude that adopting the modified call flow will give a CSFB call setup time within the range of 3G call setup time, which will make this solution more attractive for operators.

Table 1: Call Setup Time Before And After Change Including The Number Of Paging Attempt

Call	Call setup times(s)	Paging time (s)	Number of Paging attempts	First ISUP Message time (s)	Call setup after change (s)
T1	10.41	1.37	1	0.68	8.36
T2	11.05	3.128	2	0.7	7.23
T3	13.13	3.119	2	1.7	8.31
T4	11.96	3.127	2	1.5	7.34
T5	13.04	3.109	2	2.19	7.74

Figure 3. Information Flow For A Mobile Terminating Roaming Retry Call After Successful Retrieval Of Routing Information (3GPP 2012)

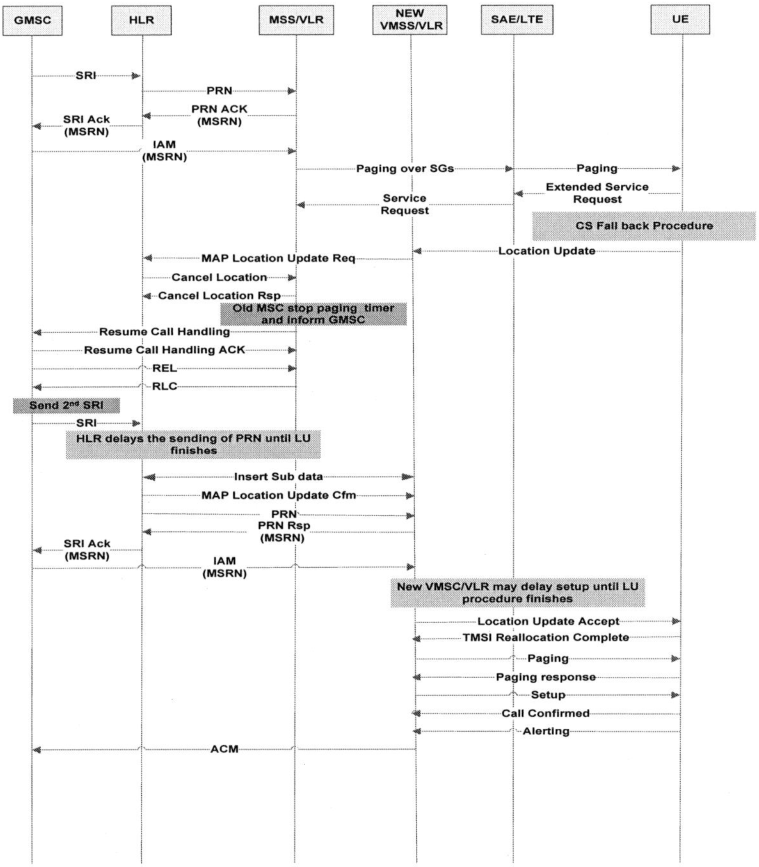

Figure 4: Successful MT Call Flow Using Proxy MSS Solution

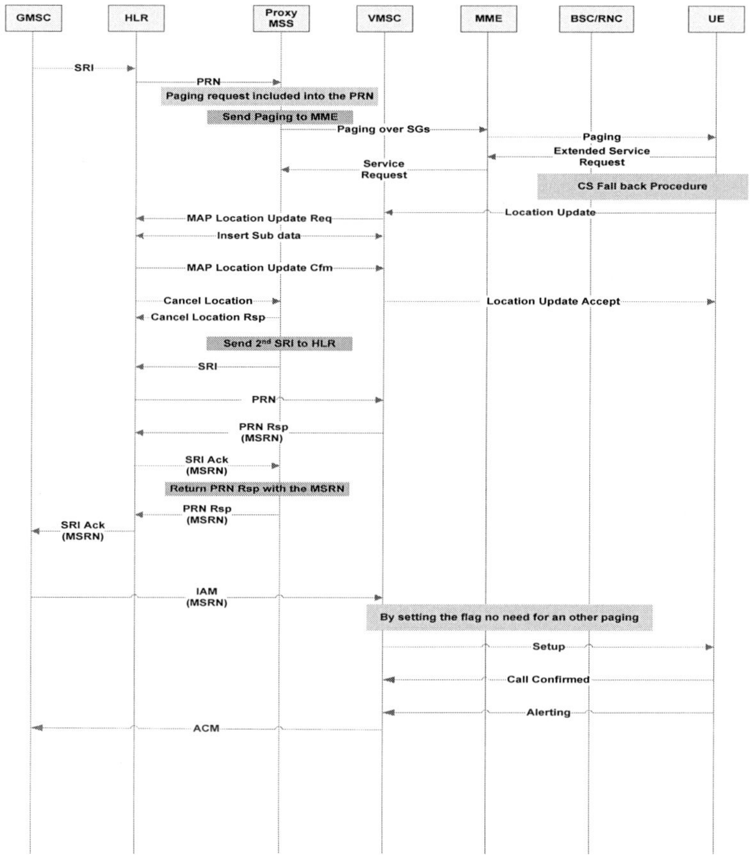

Figure 5: Comparing Call Setup Time Before And After Change

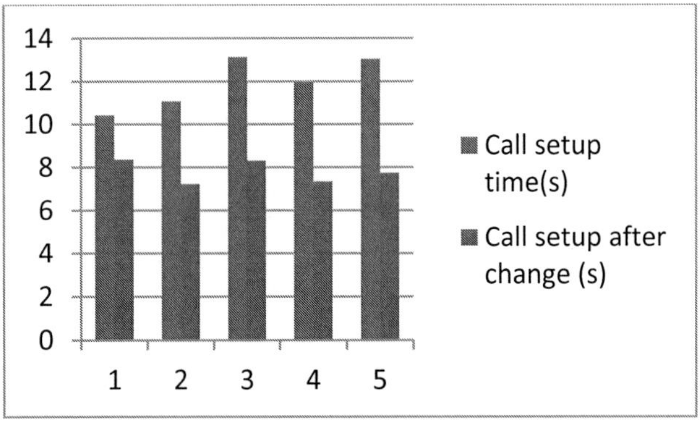

Summary

In this paper we presented, for operators ready to add voice services on their LTE networks, a smart and simple method of implementing a CSFB solution to eliminate all the confusion for a smooth voice-over LTE deployment. This solution will allow operators to announce the availability of voice services by adopting a quick and small change on their networks, which will result in minimal increased costs at a time when the majority of LTE operators are forced to wait for a viable handset supporting the solution based on SRVCC. After providing an intermediate solution to offer a comfortable situation for operators waiting for the next step in the LTE voice roadmap, a step where high definition (HD) voice and many other luxurious services will be possible, we have recommended a modified call flow to reduce the call setup time and provide a comfortable user experience, even on the most-impacted use case including the modifications used to simplify CSFB implementation. The interim CSFB solution might be used even after implementation of the SRVCC solution with roamers not yet supporting the SRVCC solution on their home networks.

References

3GPP TS 23.272. (2012) "Circuit Switched (CS) fallback in Evolved Packet System (EPS)", V8.11.0, pp. 6, 8, 9, 32 available at <http://www.arib.or.jp/english/html/overview/doc/STD-T63v9_30/5_Appendix/Rel8/23/23272-8b0.pdf>.

Asmar, T. Kneeshaw, W. (2012) "Qualcomm Chipset Powers First Successful VoIP-over-LTE Call with Single Radio Voice Call Continuity", *Qualcomm media*, viewed on 12 August 2012, <http://www.qualcomm.com/media/releases/2012/02/02/qualcomm-chipset-powers-first-successful-voip-over-lte-call-single-radio-v>.

Gompa, N. (2012) "First LTE to WCDMA voice call handover completed by Qualcomm and Ericsson", *EXTREMETECH*, viewed on 14 August 2012, <http://www.extremetech.com/electronics/117107-first-lte-to-wcdma-voice-call-handover-completed-by-qualcomm-and-ericsson>.

Jouihri, Y. (2012) "Solutions Comparison towards Voice Services Implementation for Operators Starting LTE Deployment", *Communications and Network*, Vol.4, No.2, pp. 122-128.

Kineto Wireless. (2009) "A critical examination of CSFB as a solution for providing Voice-on-LTE", *Disruptive Analysis*, viewed on 15 July 2012, <http://www.disruptive-analysis.com>.

Nokia Siemens Networks. (2012) "Feature 1914: CS Fallback in EPS for MSC Server", *Nokia Siemens Networks DX MSC/MSS /DX HLR, Feature Documentation*, pp. 9,54.

Qualcomm. (2012), *Qualcomm media*, viewed on 15 August 2012, <http://www.qualcomm.com/media/documents/files/circuit-switched-fallback-the-first-phase-of-voice-evolution-for-mobile-lte-devices.pdf>.

Qunhui, C. (2011) "Evolution and deployment of VoLTE", *Huawei Communicate,* Vol. 61, viewed on 05 October 2011, <www.huawei.com/en/static/hw-094164.pdf>.

CHAPTER FIFTEEN

AN ARCHITECTURE FRAMEWORK TO DESIGN A HEALTHCARE INFORMATION SYSTEM

SIDI MOHAMED O. MOULAYE ABDELLAHI, MOHAMED T. KIMOUR AND MBAYE SENE

Introduction

Healthcare information systems (HIS) are important to both individuals and governments, and their growing costs to the economy have contributed to make such systems an important area of research and development (Patrick et al. 2008). In addition to the network embedded devices in clinical and diagnostics equipment, HIS offer a means to capture, store, process, and communicate timely information to decision makers for better coordination of healthcare at both the individual and population levels. Advances in web technology ensure remote and constant access to a patient's information and enables new directions in development of HIS (Kirovski et al. 2007, Ko, Lee and Lee 2008).

HIS should be designed to support both patients and their healthcare providers in the process of monitoring and treatment of patients (Sidi-Mohamed, Kimour, and Sene 2012). However, little work in integrating embedded devices with internet applications for the support of patients has been done to date (Lee et al. 2008, Chenhui, Huilong, and Xudong 2008), (Spasov and Petrova 2008).

In this paper, we present an architectural framework for the design of health information systems to support the monitoring of the conditions of patients with chronic diseases and the remote monitoring of the patient. This design would establish a virtual link between patients and their caregivers to support the timely delivery of health-related data (Al-hakim 2007). In fact, our framework extends our previous work in Sidi-Mohamed, Kimour, and Sene (2012) by identifying the functional components at the side of the patient and the clinician. We demonstrate how to identify subsystems, and for each subsystem, we demonstrate how to identify components.

The architecture of such a system should incorporate a local network at the patient's residence that wirelessly connects various sensors, actuators, and a local server. The latter is also wirelessly connected to a web portal database server, to which some important measurements of the body of the patient are continuously delivered. Situated at the monitoring center, the central server should be accessible by clinicians such as doctors, nurses, etc.

The HIS continuously collects a wide range of physiological data of patients (blood pressure, heart rate, etc.), their physical activity data (sleeping, exercising, etc.), as well as environmental conditions including temperature and humidity. The HIS provides more accurate data from the continuous monitoring of patients in their natural environment for clinicians (Carvalho, Murta, and Loques 2012).

Consequently, the HIS should be able to adapt to changes in the patient needs and preferences in the residential environment by adding new resources, e.g., sensors, devices. The patients'needs can be changed as the health problems and conditions change. This may require a new appropriate physiological sensor and new software resources. (Carvalho, Murta, and Loques 2012)

To efficiently deal with the aforementioned characteristics of the HIS, our architectural framework should consider the best practices and guidelines, inspired from the most notable web developments processes such as WebML (Ceri, Fraternali, and Bongio 2000) and UWE (Koch and Kraus 2002), but also from methods dedicated to distributed systems (Gomaa 2000).

The remainder of this paper is organized as follows:

- Section 2 presents an overview of the use of embedded systems in healthcare.
- Section 3 discusses our architectural framework for the design of a HIS.
- Section 4 concludes and provides direction for future research.

Embedded Systems in Healthcare

Embedded systems in the HIS perform some specific tasks. They have greatly evolved in aspects such as mobility, wireless communication, integration with internet, computing performance, widely enhanced security, etc. They offer new applications in various domains. One of the most important domains is the healthcare system.

For instance, as an embedded system, the mobile phone is more and more used as a valuable tool for patient monitoring. In fact, there have been many advances in mobile technology, such as serial ports and Internet connections. Therefore, the phones can interact with biosensor devices, such as patient biosensors, and transmit vital signals through Internet protocols, such as TCP/IP.

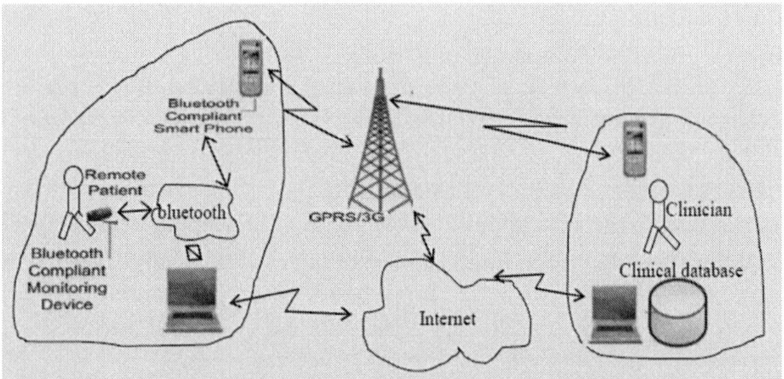

Figure 1: A topological view of the healthcare information system.

This new, mobile communication technology helps patients perform a wide variety of medical tasks without the assistance of trained medical personnel. For example, diabetic patients can check their blood glucose level several times a day and adjust the insulin dose appropriately. They may also check crucial medical parameters such as blood pressure. This medical data can be sent over the cell phone to the local physician's office (WMA 1992).

A care plan guides the patient regarding measurements that must be done, e.g., blood pressure, medications that should be taken, and other personalized recommendations according to the treatment (Istepanian, Jovanov, and Y.T. 2004, Loques and Sztajnberg 2010, Park and Jayaraman 2003).

The Architecture Framework

Health information systems are increasingly complex and dynamic. The design activity provides the architectural description of the system compliant with the used component model. In the design of such systems, the current best practices dictate the use of hierarchical distributed architectural models. The entire system is considered as a collection of concurrent subsystems, possibly communicating with each other.

Our design flow is composed of four steps. First, we outline an overview of the system's objectives, missions, and the technical infrastructure (embedded devices and networks). Second, we decompose the system into subsystems. At this step, end-to-end requirements are depicted, especially those related to timing requirements. Third, we decompose each subsystem into functional components. Finally, for each subsystem, we define its components (either hardware or software). In doing so, our architectural framework advocates to perform two levels of structural decomposition: the subsystem level and the component level.

System Overview

The HIS should constantly monitor the patients' conditions and remind them to perform needed tasks, such as blood pressure and glucose level measurements. It should communicate such measurements to a medical tracking center and eventually remotely notify the doctor about critical situations (Figure 1).

HIS should help patients at home (especially the elderly ones) when they have to manage specific medical tasks. HIS must keep track of their regimes of medications and doctor appointments, as well as check parameters such as blood sugar or blood pressure. Medical staff (doctors, nurses, and physicians) has the opportunity to remotely monitor the patients' daily activities and adjust his/her care plan based on the treatment progress. For the patients, this can mean fewer visits to the doctor and shorter periods of hospitalization.

A home wireless local area network (WLAN) is needed in a home healthcare system. The communication is usually based on one of the wireless standards from IEEE 802.15 wireless network family (Istepanian, Jovanov, and Y.T. 2004). The warning messages to the patient should be shown in one or more display devices such as TV, cellular phone, tablet, PC, etc.

Subsystems Identification

Subsystems are the system parts' design or implementation units that can be developed independently and separately distributed (Kimour, Boudour, and Ghanemi 2012).

The subsystem view specifies a coarse-grained component model used for modeling independently distributed components with complex functionality (Koch and Kraus 2002, Kimour, Boudour, and Ghanemi 2012). It is based on two important concepts: separation of concerns and distribution. Separation of concerns promotes system comprehension and reuse, hence reducing cost and development delay.

Thus, it is evident that our HIS distinguishes two topologically distributed units, which are connected by the Internet. The first one is at the patient side, the Local Patient Monitoring (LPM) subsystem, and the other is at the healthcare center side, the Healthcare Center (HC) subsystem.

The LPM subsystem in Figure 2 aims to guide the patient on when to measure blood pressure and when to take medication, based on the data collected locally. It also aims to inform them of other personalized recommendations according to the treatment (Istepanian, Jovanov, and Y.T. 2004, Kimour, Boudour, and Ghanemi 2012).

Using appropriate mobile devices (sensors), a patient's measurements, such as glucose level, weight, and blood pressure, may be constantly registered to the home gateway. A PC, PDA, or even a smart phone may be such gateway that automatically uploads (via the Internet infrastructure) a patient's measurements to a server (healthcare center) for access by healthcare providers.

Figure 2: The LPM subsystem. Dashed elements correspond to optional

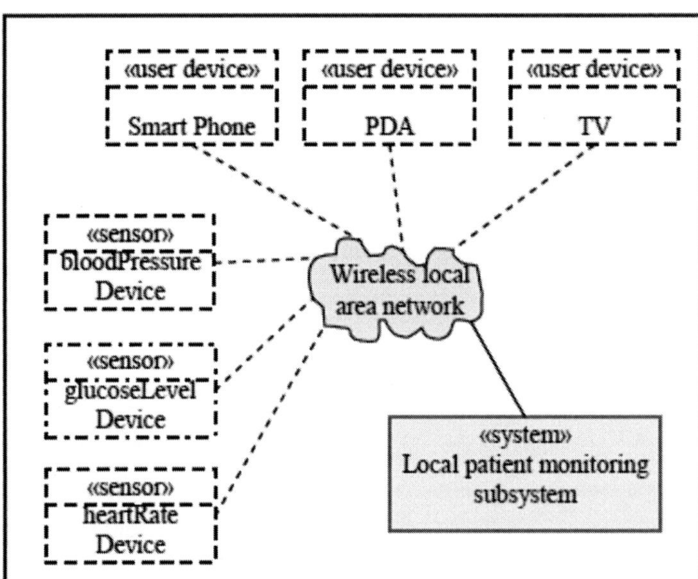

The LPM and HC interacts through the gateway to exchange patients' medical data collected from wireless sensor networks of blood pressure devices, glucose level devices, and heart rate monitoring devices. There are many systems that allow the automatic and complex telemonitoring of persons with chronic illnesses, elderly people, and those with high medical risks. One such example is TELEMON, the integrated system for real time telemonitoring of patients and elderly people (Costin et al. 2009). The TELEMON system consists of subsystems that are similar to ours, and it has similar functional characteristics. Costin et al. describe its local subsystem this way (2009).

> The Local Subsystem for the home monitoring of the patient (Figure 2) is built around a personal computer (PC) which receives data from the patient. The data are provided by a personal sensors/transducers network in form of medical parameters such as ECG, SpO2, blood pressure, body temperature a.s.o. [16]. The outputs of the network are connected to a radio module that sends the data to the PC. A very important feature of the network and also of the radio module is their low power consumption. Another component of a local subsystem is the sensors/ actuators for domotics. The domotic sensors collect data from different points of the house sends them to the PC. The PC takes the decisions and sends commands to the actuators such as: light adjustment by controlling the windows, by controlling the patient access at doors level, by controlling the room's temperature, by controlling the audio-video receivers a.s.o. The Local Subsystem allows the video monitoring of the patient through a video camera connected to the PC. The PC records and stores images of the patient during the acquisition of the medical parameters, process these images and can send to the database server images and also processing results (e.g., alarms). The PC is equipped with a radio module that receives data from the patient's personal network, runs the monitoring algorithms, activates the alarms if the medical parameters are above the limits and sends the results to the database server (pp. 4-5).

Component Identification

To be able to build software systems out of reusable components, we need a component model. A component model is mainly based on the concepts of component, port, interface, and a connector (Kimour, Boudour, and Ghanemi 2012).

For each subsystem, we distinguish among five types of components.

- Data. This component contains persistent information recording patient status of situation (Gomaa 2000). A data component is defined in order to handle persistent data such as patient health status. Usually, this component is defined as a database.

- Control. This component handles interactions between components and models cohesive application logics (algorithmic). The control component is defined to handle specific computing inherent to coherently related business rules.
- Interface. An interface component is defined for each input or output device, such as a sensor, actuator, display device, etc.
- Variability.
- Coordination. A coordination component defines the entire functional computing of the subsystem.

To handle the communication issue, we define a communication component which will be assigned the task to interface with the network. In the following, we show how to determine components for the LPM subsystem (Figure 3).

Indeed, we define an interface component to each input/output device. For instance, to handle interactions with the devices depicted in Figure 2, we identify the following interface components: blood pressure, heart rate, glucose level, Smart Phone, PDA, and TV.

We identify a data type component as a "patient status component" to handle persistent data and historical states for the patient.

To identify control components, we lay on the distinguished expertise criterion. That is, we define components for heart rate control, blood pressure control, and glucose level control.

Finally, to enable adaptability at run-time, we allow customization through the use of an adaptability component. Indeed, such a component will handle changes in the patient's health situation during the system operation that may lead to modifications in resource configuration, such as medical sensors, or other input/output devices.

In case of the availability of various devices at home (TV, PDA, tablet, etc.), they could be selected and bound at runtime. This selection is based on context rules that should be defined at deployment time.

To do this, we represent adaptability at the architectural level that we associate with a decision table to formalize context rules of selection at run- time. Therefore, in terms of the product line architecture, these devices can be defined as variabilities whose architectural elements can be bound at deployment time or at runtime (Carvalho, Murta, and Loques 2012).

Figure 3: The compenent model of the local patient monitoring subsystem

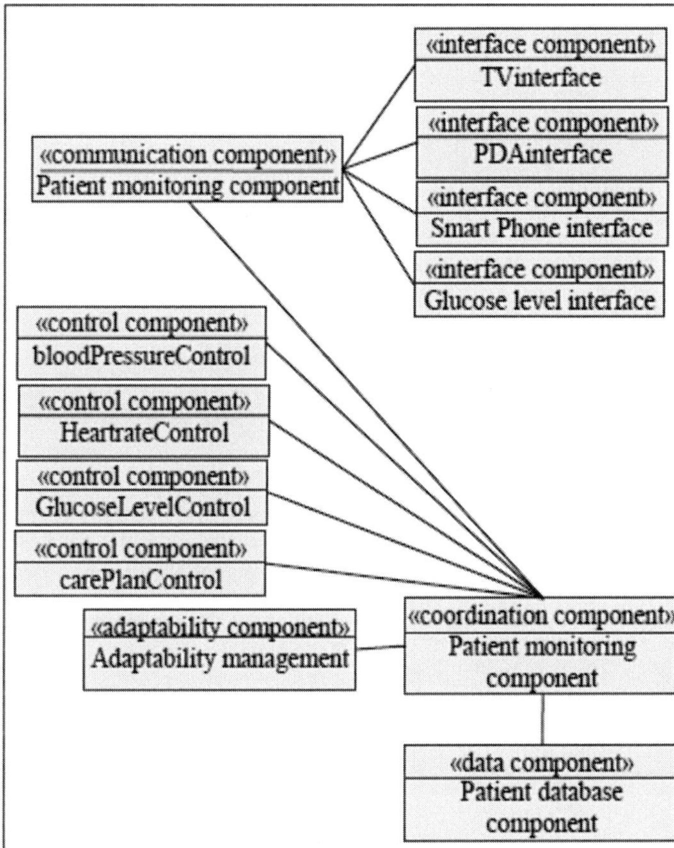

In our approach, adaptabilities are described as architectural variations in elements and conditions, which determine the architectural adaptation actions.

In the adaptability component, the decision table specifies that the system selects the device only if the patient is located by a location service of the local monitoring component that is in the same room of the device.

Due to limitations in space, we only decompose the subsystem of the local patient monitoring (HC). The remaining ones could be developed in the same manner. Indeed, the LPM subsystem will be decomposed according to the separation of concerns principles, leading to the five types of components in embedded systems, which are mentioned previously.

Conclusions and Future Work

In this paper we have presented an architectural framework for healthcare information system design. Such a framework deals with the integration of modern technologies in theWeb and mobility that improve the presentation and delivery of healthcare.

It guides the designer in building a healthcare information system that allows the patients at home to be responsible for their own health and to contribute in managing the health delivery process. Using such a system, physicians will be provided with more accurate data from continuous monitoring of patients in their natural environment. In doing so, doctors will be able to give more accurate prescribes. In future works, we plan to use environmental support for system validation and automatic code generation for a target platform.

References

Al-hakim, Latif. 2007. *Web Mobile-Based Applications for Healthcare Management*. Hershey, PA: IRM Press.

Carvalho, S.T., L. Murta, and O. Loques. 2012. Variabilities as first-class elements in product line architectures of homecare systems. Paper read at 2012 4th International Workshop on Software Engineering in Health Care (SEHC), 4-5 June, at Zurich.

Ceri, Stefano, Piero Fraternali, and Aldo Bongio. 2000. "Web Modeling Language (WebML): a modeling language for designing Web sites." *Computer Networks* no. 33 (1-6): 137-157.

Chenhui, Zhao, Duan Huilong, and Lu Xudong. 2008. "An integration Approach of Healthcare Information System." In *Proceedings of IEEE Conference on Biomedical Engineering and Informatics (BMEI)*, 606-609.

Costin, Hariton, Vlad Cehan, Cristian Rotariu, Octavia Morancea, Victor Felea, Ioana Alexa, Gladiola Andruseac, and Ciprian Costin. 2009. "TELEMON – A Complex System for Real Time Telemonitoring of Chronic Patients and Elderly People." In *Proceedings of the 4th European Conference of the International Federation for Medical and Biological Engineering (IFMBE)*, 1002-1005.

Gomaa, H. 2000. *Designing Concurrent, Distributed, and Real- Time Applications with UML*: Addison Wesley.

Istepanian, R., E. Jovanov, and Zhang Y.T. 2004. "Beyond Seamless Mobility and Global Wireless Health-Care Connectivity."

IEEETransactions on Information Technology in Biomedicine no. 8 (4): 405-441.

Kimour, M. T., R. Boudour, and S. Ghanemi. 2012. Closing the gap between the requirements specification and the design model for embedded systems. Paper read at Second International Symposium on Modelling and Implementation of Complex Systems (MISC 2012), May at Constantine, Algeria.

Kirovski, D., N. Oliver, M. Sinclair, and D. Tan. 2007. Health- OS: A Position Paper. Paper read at the 1st ACM SIGMOBILE International Workshop on Systems and Networking Support for Healthcare and Assisted Living Environments, June 11, at San Juan, Puerto Rico.

Ko, E.J., H.J. Lee, and L.W. Lee. 2008. "Ontology and CDSS based Intelligent Health Data Management in HealthCare Server." In *Proceedings of World Academy of Science Engineering and Technology*, 119-123. Vienna, Austria.

Koch, Nora, and Andreas Kraus. 2002. "The expressive power of UML-based web engineering." In *Proceedings of the Second International Workshop on Web-oriented Software Technology (IWWOST02)*, 105-119. Springer.

Lee, H., S. Lee, K. Ha, H. Jang, W. Chung, J. Kim, Y. Chang, and D. Yoo. 2008. "Ubiquitous healthcare service using Zigbee and mobile phone for elderly patients." *Int.Journal of Medical Informatics* (78): 193-198.

Loques, Orlando, and Alexandre Sztajnberg. 2010. "Adaptation issues in software architectures of remote health care systems." In *SEHC '10 Proceedings of the 2010 ICSE Workshop on Software Engineering in Health Care*, 24-28 New York, NY: ACM

M. Eslami and M. van Sinderen, "Flexible Home Care Automation Adapting to the Personal and Evolving Needs and Situations of the Patient", in 3rd Pervasive Health, Los Alamitos, 2009, pp. 1-2.

Park, S., and S. Jayaraman. 2003. "Enhancing the Quality of Life Through Wearable Technology." *IEEE Engineering in Medicine and Biology* no. 22 (3): 41-48.

Patrick, K., W. G. Griswold, F. Raab, and S. S. Intille. 2008. "Health and the mobile phone." *Amer. J. Prev. Med.* no. 35 (2): 177-181.

Sidi-Mohamed, M. A., M. T. Kimour, and M. Sene. 2012. Designing Web Imbedded Systems for healthcare. Paper read at International Conference on Software Engineering and Technologies December 15-17, at Hammamet, Tunisia.

Spasov, Grisha, and Galidia Petrova. 2008. Web-based Personal Health Systems – Models and Specifics. Paper read at International Scientific Conference Computer Science, November, at Kavala, Greece.

WMA, (World Medical Association). 2013. *Statement on Home Medical Monitoring, Tele-Medicine and Medical Ethics* 1992 [cited October 5 2013]. Available from http://www1.umn.edu/humanrts/instree/home monitoring.html.

PART III

MISCELLANEOUS TOPICS

CHAPTER SIXTEEN

A NEW APPROACH FOR HUMAN ASSESSMENT OF ONTOLOGIES

LEILA ZEMMOUCHI-GHOMARI
AND ABDESSAMED RÉDA GHOMARI

Introduction

Ontology evaluation, from logical and empirical perspective, has been recognized as a critical phase in the ontological engineering research field (Eschenbach and Gruninger 2008). Most well-known ontology development methodologies include evaluation (Fernandez-Lopez et al. 1997, Sure and Studer 2002, Suárez –Figueroa et al. 2012), either at the end or throughout the whole ontology development process (Lovrenčić and Čubrilo 2008). However, the profusion of methodologies has made it difficult for researchers to develop generic approaches for evaluating ontological quality that are independent of any specific methodology (Hulden 2007).

On the other hand, proliferation of low-quality ontologies is problematic. Agents that exploit ontologies that encode incomplete, inconsistent, inaccurate, or ambiguous knowledge cannot perform tasks successfully. Indeed, poor-quality ontology can reduce efficiency and effectiveness by providing agents with poor information (Burton-Jones et al. 2005).

Gomez-Perez (2004) introduced the two paradigms for describing ontology evaluation: ontology verification, and ontology validation. Ontology verification deals with building the ontology correctly, while ontology validation, the focus of this paper, refers to the correspondence between the semantics of the model and the real world for which the ontology was designed.

Actually, ontology validation approaches can be classified according to several perspectives, namely:

- Comparison with a gold standard or a reference ontology (Zemmouchi-Ghomari and Ghomari 2009);
- Comparison with a source of data (collection of documents that covers the ontology domain);
- Human assessment (ontology developer, domain expert and end-user); and
- Application based-ontology assessment (Brank et al. 2005).

Beyond this classification, all of these approaches require the close cooperation of domain experts. According to Gomez-Perez (2004), ontological validation often cannot be performed automatically.

Besides, ontology itself can be defined as a social artifact, since it specifies shared agreement of a group of stakeholders with regard to the conceptualization specified in the ontology. In other words, consensus is measured through the proportion of agreement that domain experts have with respect to ontology elements (Gangemi et al. 2005).

The objective of this paper is to propose a methodological baseline for ontology assessment by humans (domain experts and end-users) via a questionnaire, intended to support a component-based evaluation. Actually, this type of evaluation is a glass box evaluation. This is in contrast to a black box (task based) evaluation that usually is performed with ontologies integrated in applications with specific tasks.

This paper is organized in the following way. Section II describes related work or ontology human assessment as reported in the literature. Next we present the proposed approach in Section III. Section IV is dedicated to describing our specific case study. i.e., evaluation of Higher Education Reference Ontology (HERO)[1] via an online questionnaire. Conclusion and perspectives are presented in Section V.

Related Work

Human assessment of ontologies, as reported in the literature, fits into the realm of ontology verification. This work is intended to detect mistakes and inconsistencies that occur with human modeling.

In fact, human assessment against a set of criteria had been used by Obrst and his colleagues (2007) in a series of studies of ontologies and terminologies in biomedicine. The principles are derived from common sense, such as: provide clear documentation and use terms in a consistent way.

---

[1] http://sourceforge.net/projects/heronto/?source=directory.

For example in the thesaurus evaluation completed by the National Cancer Institute (NCI), an analysis of the thesaurus was performed in order to assess its conformity with principles of good practice in terminology and ontology design (Ceusters et al. 2005). The authors found many mistakes concerning the formation of terms, the underlying knowledge representation system, and missing or inappropriately allocated informal and formal definitions.

Likewise, in the Systematized Nomenclature of Medicine (SNOMED) evaluation (Ceusters et al. 2004), some mistakes seem to come from human error that happens with manual modeling, for example, improper assignment of IS-A relationships and non-IS-A relationships, shifts in terms meaning, and redundancy in concepts.

It appears that "human assessment" in this instance is not expert nor is it the end-user assessment as targeted by this study. The scope of this paper fits into the field of ontology validation, which relates a system to its end users' implicit requirements. Actually, evaluation by domain experts or end-users is difficult to carry out and has not been tackled often by researchers. That is because some quality attributes judged by this type of user, such as clarity, relevance, and accuracy, can be difficult to evaluate as they may not be easily quantifiable.

In what follows, we are going to focus mainly on domain expert and, to a lesser extent, end-users in the assessment of ontologies. The experts are, of course, more proficient in evaluating the knowledge encrypted in the ontology and its degree of correspondence to that part of reality that it is designed to represent.

Ontology Human Assessment Approach

Ontology editors cannot play well enough in ontology evaluation because they must be downloaded and installed to be run. In this paper we would like to introduce a web-based evaluation system composed of an online questionnaire through which evaluators can rate the knowledge encoded in the ontology.

Since the human language is considered to be a primary means of knowledge transfer, the selected validation technique is a natural language questionnaire.

The assessors, in the context of academic domain, include researchers, teachers, administrators and students (current and former, or alumni).

The proposed approach is composed of five steps (Figure 1). The approach begins with a candidate ontology and finishes with an updated ontology according to the recommendations of the assessors:

1) Derivation of questionnaire from ontology components;
2) Aggregation of survey results and free comments/suggestions;
3) Analysis and synthesis of obtained results;
4) Update of the questionnaire with regard to experts' view points; and
5) Update of the ontology according to derived knowledge from refined results.

Figure 1: Architecture of Proposed Ontology Validation System Through Questionnaire

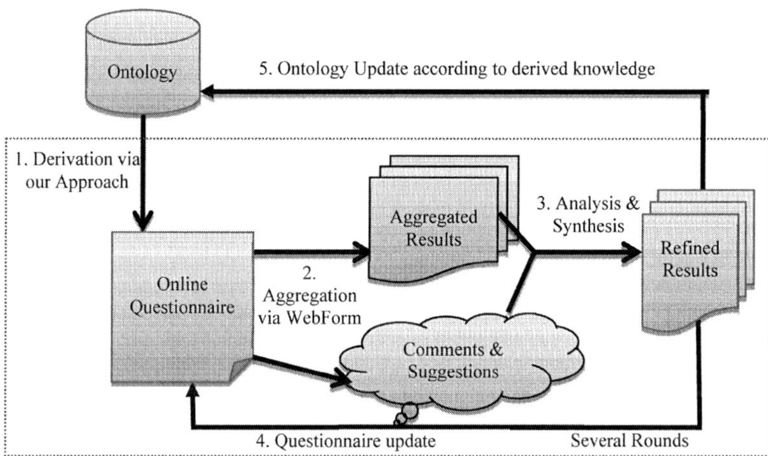

In this paper we are going to focus first on step 1 of the proposed approach. Step 2 is performed automatically by a web form module (like drupal webform module, available at: http://drupal.org/project/webform). After commenting on the first step, we consequently will provide suggestions for how to best deal with the questionnaire feedback in step 3 and step 4.

Step 1: Derivation of Questionnaire from Ontology

The questionnaire is intended to represent the most relevant constituents of the ontology in an informal way (natural language) in order to be assessed by domain experts who are not necessarily proficient in web ontology language (RDF or OWL).

In fact, an ontology can be evaluated merely on the basis of its components since every ontology is necessarily composed of similar elements, i.e., a hierarchical organization of concepts described by attributes, linked by relations, and interpreted according to axioms.

Figure 2 shows the direct relationships among ontology concepts: ontology, class (class and concept are used interchangeably in this paper), attribute, instance, relation, axiom, language, methodology, and tool. These concepts are used here as defined in the Reference Ontology (Lozano-Tello and Gomez-Pérez 2004).

Figure 2: Concepts of Reference Ontology, adapted from (Lozano-Tello and Gomez-Pérez 2004)

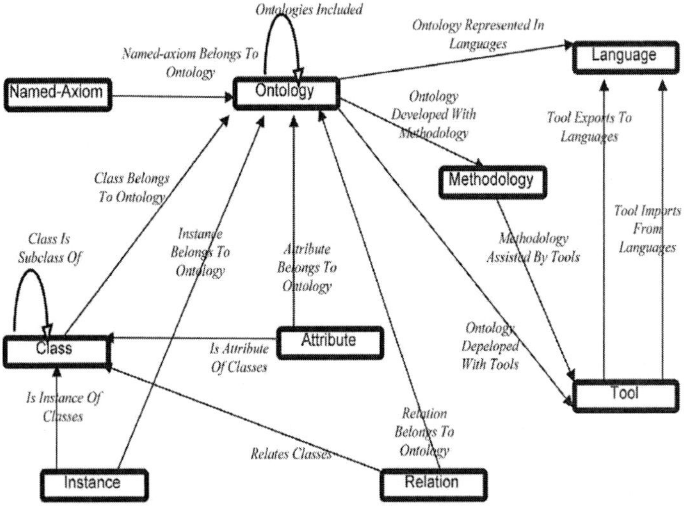

Since ontology evaluators in the scope of this paper are not ontology developers, we do not take into account the methodological aspects of building ontologies illustrated by figure 2, namely, language, methodology, and tool. Also, the concept of instance is not relevant when analyzing a reference ontology (Zemmouchi-Ghomari and Ghomari 2009) because this ontology type is intended to describe generic aspects of a given domain and instance level is not as important as the other levels in this context.

As ontology is a fairly complex structure, it is more practical to focus on the evaluation of different levels of the ontology separately rather than trying to evaluate the ontology directly as a whole (Brank et al. 2005). That is why we decided to divide the survey into several parts, namely:

- **Hierarchical ontology levels** (ontology depth): The structure of an ontology is a graph. The most widely explored measures used on ontologies are structural measures (Völker et al. 2008). Graph measures are applied to the complete or partial graph describing the ontology. An example of an extensively investigated subgraph would be the one consisting only of edges with the name *rdfs:subClassOf* and the nodes connected by these edges (i.e. the explicit class hierarchy).
- **Axioms:** This set of axioms provides a declarative specification for the various definitions and constraints on the ontology (Fox et al. 1996).
- **Relations between concepts:** or so called horizontal relations (Hulden 2007). In fact, "an ontology that contains only inheritance relationships usually conveys less information than an ontology that contains a diverse set of relationships" (Tartir et al. 2010).
- **Descriptive attributes of concepts**: or properties of the concept that describe the concept itself.

Actually, we have to confirm whether the foregoing components comply with the expertise of one or more evaluators. Does the ontology capture and correctly represent aspects of the real world? Does the ontology include irrelevant components with regard to the domain to be covered (Gomez-Perez 2004)?

The results of applying a particular method will provide indicators for how well one or more criteria are met. Inversely, "*Criteria provide justifications for the methods*" (Völker et al. 2008). In order to propose possible answers to the questions included in our survey, we rely on principles that are mandatory for good-quality ontologies, defined by (Gomez-Perez 2004, Obrst et al. 2007, Gruber 1995, Burton-Jones et al. 2005, Fox et al. 1996). This approach is summarized in Table 1.

Table 1: Mapping between criteria and possible answers to questionnaire

Criterion	Materialization of this criterion in possible answers	Answers' Location(s)
Clarity (Gomez-Perez 2004, Fox et al. 1996): verify if "the ontology is easily understood by the users so that it can be consistently applied and interpreted across the domain of interest?"	- *Not Clear*	- Classes' validation - Axioms' validation - Relations' validation - Attributes' validation
Lawfulness (Burton-Jones et al. 2005): or Syntactic Quality: verify if the knowledge described by the ontology is encoded with meaningful terms. This is achieved by checking to assure that the words used by the ontology are appropriate.	- *Right but another term would be more appropriate*	- Relations' validation
	- *Relevant but used terms are not appropriate*	- Attributes' validation
Accuracy (Gomez-Perez 2004, Burton-Jones et al. 2005): determine whether the claims an ontology makes are 'right' or 'wrong.'	- *Right* - *Wrong*	- Classes' validation - Relations' validation
	- *Always* - *Sometimes* - *Never*	- Axioms' validation
Relevance (Burton-Jones et al. 2005) or Pragmatic quality: verify whether the ontology satisfies ontology requirements or not.	- *Relevant* - *Not really relevant* - *Not relevant at all*	- Attributes' validation

Selection of assessed ontology elements with respect to good quality ontology criteria has been guided by these principles:

Clarity of all ontology components is a fundamental criterion for achieving knowledge-sharing among members of a given community;

Lawfulness can be extended to classes and axioms provided that classes and axioms definitions[2] will be available when requested in order to avoid ambiguity issues. (Unlike relations who are defined by classes they link to each other and attributes which are related to specific classes.) In other words, relations' and attributes' contexts are already defined;

Accuracy is more subtle when validating axioms' statements, that is why we proposed more than two Boolean options;

Relevance of proposed attributes must be checked, because most of them are proposed on the basis of team developer knowledge background in contrast to other ontology elements (classes, relations), which are directly extracted from competency questions/answers pairs specified in the first step of the ontology building-process (as required by several ontology-building methodologies). For example: date and place of birth, and attributes of class student are not likely to be derived from the ontology specification phase, although they often are included in the list of class student attributes.

Step 3 and Step 4: Analysis/Synthesis of Obtained Results and Questionnaire Update

Holsapple and Joshi (1963) proposed an evaluation method for collaborative manual ontology engineering, in which each suggestion made by one expert is evaluated by other experts. Hence, good suggestions are those that enjoy the approval of multiple experts.

One known technique based on this principle is the Delphi Method, developed primarily by Dalkey and Helmer (1963) at the Rand Corporation in the 1950s. It is a widely used and accepted method for achieving convergence of opinions concerning real-world knowledge requested from experts within certain topic areas (Hsu and Sandford 2007).

The Delphi method is well suited as a means and method for consensus-building by using a series of questionnaires to collect data. It employs multiple iterations designed to develop a consensus of opinion concerning a specific topic.

Theoretically, the Delphi method can be iterated continuously until it is determined that consensus has been achieved. However, previous experience points out that three iterations often are sufficient.

Clearly, the Delphi method can constitute a suitable candidate method in order to deal with questionnaire feedback until consensus is reached

[2] As annotations for example.

(step 3 and step 4), even if the definition of consensus is not trivial as well. According to Gomez-Perez (2004), the bigger a group that commits to an ontology, the harder it is to reach a consensus -- but also the bigger the potential benefit. Thus the status of the ontology with regard to relevant standardization bodies in the given domain is a major criterion when evaluating an ontology.

The Delphi method definitely has a role to play in obtaining experts' semantic agreement, but it has been relatively underexplored for use in ontology evaluation (Dalkey and Helmer 1963).

However, the Delphi method takes time and requires a lot of effort. Some experiments have highlighted these drawbacks, such as that reported in Booto et al. (2011) where, out of a total of 256 potential experts, 22 participated in first round of the survey with only 16 of 22 returning the completed questionnaires. The completion rate was even more disappointing in the second round, with only 16 experts participating and 13 submitting their questionnaires. That represents just 5% of initial list of experts.

In terms of time scheduling, this experience showed that, instead of two weeks for two rounds as originally planned by authors, the Delphi survey took over eight weeks.

Case Study

As mentioned in step 1, we derived a questionnaire (100 questions, at http://herontology.esi.dz/content/questionnaire-hero-ontology), from an ontology we built called HERO. This derivation is based on ontology components for the 'questions' part of the survey. We also proposed multiple choice questions (MCQ) for possible answers according to the good-quality ontology criteria described in Table 1.

The questionnaire was divided into four parts covering the five hierarchical levels of the HERO ontology, classes, axioms, relations, and attributes. Sample screenshots of this survey are presented below (Figure 3).

The answers from domain experts to the questionnaire are shown in Table 2. Even if results of questionnaire provided precious information on the accuracy, clarity, and relevance of the evaluated ontology, we found open-ended comments requested after each section of the questionnaire to be extremely useful.

Sections available after each subset of the questionnaire have been very instructive and helpful. We think that proposed MCQs stimulated domain experts to rectify, or to make more precise, some ontology statements presented previously.

Figure 3: Sample screenshots of the HERO Ontology Online Questionnaire

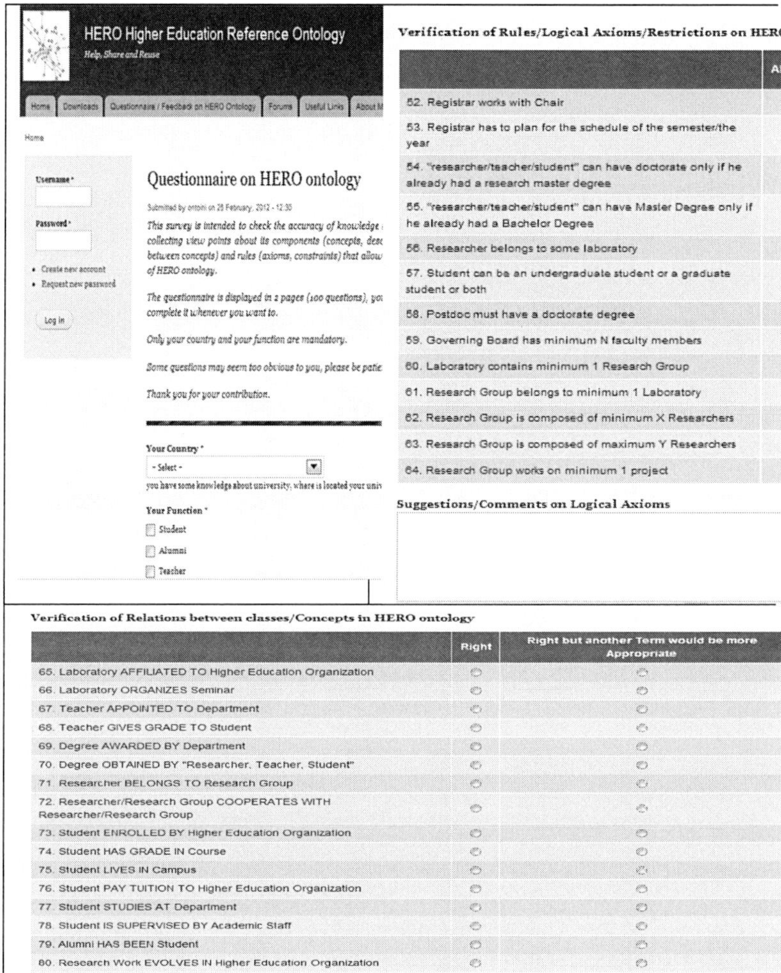

We note that the axioms part of the questionnaire (Table 2) did not achieve a majority concensus (more than 50%) because of its ambiguity or its non-concordance to the real world. This inspired us to reformulate these

axioms by using other terms, for example, or by rechecking their conformity with regard to real-world assumptions.

More generally, no part of the questionnaire obtained an absolute majority consensus, defined as more than 80% according to Booto et al. (2011). Such an absolute majority would be required for a true consensus on any given topic.

Nonetheless, the ontology has been updated according to some consensuated opinions' experts. Table 3 summarizes some recommendations and their respective consequences on evaluated ontology.

Table 2: Aggregated Results of HERO ontology questionnaire

Classes' Hierarchy levels ***51 questions***	Right: 66%	Wrong: 14%	Not Clear: 20%		
Relations ***26 questions***	Right: 56%	Right but another term would be more appropriate: 27%	Wrong: 17%	Not Clear: 0%	
Axioms ***13 questions***	Always: 48%	Sometimes: 25%	Never: 0%	Not Clear: 27%	
Attributes ***10 questions***	Relevant: 52%	Not really relevant: 17%	Relevant but another term would be more appropriate: 25%	Not relevant at all: 5%	Not Clear: 0%

The update of the questionnaire based on feedback from experts (step 4), has led to some changes, such as describing all ontology subsumption relations in the questionnaire by "subclass of" which is a more generic and unifying trem than "is a kind of," which or "is a" (both wordings having been a source of controversy among experts).

In general, while we think that results obtained provide indicative but not definitive evidence for assessing the quality of any ontology, other rounds with experts must be performed to realize a more reliable judgment.

Table 3: Ontology update according to experts' recommendations (excerpt)

Recommendation	Action on the ontology
A student can be enrolled in an undergraduate program and in a graduate program at the same time; while we have declared undergraduate student class and graduate student class as disjoint[3] classes	We have to remove incompatibility axiom between graduate and undergraduate students because there exists some students who overlap the two categories
Postgraduate student class has been identified as missing class in the ontology	Add a new class «postgraduate» which is a sub class of «student»
Some terms used to describe some classes were not agreed upon by all experts.	Add labels to some classes, in order to avoid ambiguity: «Research Group» or «Research Team», «Technical Report» or «Research Report»

Since obtaining experts' consensus is a major challenge, we think that the DELPHI method can contribute significantly to obtaining agreement with regard to knowledge encoded in the ontology. This work falls within perspectives of the current proposal.

Conclusion

Because the Semantic Web relies on ontologies, the lack of high-quality ontologies remains a significant barrier to its growth and feasibility.

In fact, quality of ontologies influences greatly their potential usability and reusability; hence the need of ontology evaluation approaches.

The purpose of this position paper was to propose a methodological baseline for human assessment of ontologies and to carry out a practical case study for its applicability, rather than to provide a ready-to-use toolset.

In brief, the proposed approach is composed of five steps, namely:

1. Derivation of questionnaire from ontology elements
2. Aggregation of survey results

[3] No common instances.

3. Analysis and Synthesis of obtained results
4. Update of the questionnaire according to experts' view points

Update of the ontology by taking into account derived knowledge from refined results

Obviously, this approach has several limitations, and the reasons are twofold. On the one hand, we have not yet fully exploited the potential of the underlying principles of good-quality ontology presented in this paper. On the other hand much more experience is needed about the practical usage of these guidelines.

Finally, no automatized method will ever suffice in or of itself (Gangemi et al. 2005) and intellectual judgment always will be needed, even if it makes evaluation more difficult, more subjective, less complete, and more time-consuming.

Acknowledgment

We are grateful to the domain experts who accepted the invitation to answer to our online questionnaire. We thank them for their constructive responses and helpful comments. Deserving special thanks are Professor Nathalie Aussenac-Gilles (IRIT, Paul Sabatier University, France), Professor Houari Sahraoui (Montreal University Canada) and Dr Abdelaziz Khadraoui (University of Geneva, Switzerland).

References

Booto, J. P. Bernard, P. and Plaisent, M. (2011) "Consensus par la méthode Delphi sur les concepts clés des capacités organisationnelles spécifiques de la gestion des connaissances", *Recherches Qualitatives*, Vol. 29, No 3, pp. 168–192

Brank, J. Grobelnik, M. and Mladenic, D. "A survey of ontology evaluation techniques", paper presented at Conference on Data Mining and Data Warehouses (SiKDD), 2005.

Burton-Jones, A. Storey, V. Sugumaran, V. (2005) "A semiotic metrics suite for assessing the quality of ontologies", *Data and Knowledge Engineering*, Vol. 55, No. 1, pp. 84–102.

Ceusters, W. Barry, S. Anand, K. and Christoffel, D. "Ontology-based error detection" Paper presented at SNOMED-CT, Medinfo, Amsterdam, pp. 482–486, 2004.

Ceusters, W. Smith, B. and Goldberg, L. (2005) "A terminological and ontological analysis of the NCI thesaurus", *Methods Inf Med*, Vol. 44, pp. 498–507.

Eschenbach, C. and Gruninger, M. "Formal Ontology in Information Systems", paper pesented at the fifth Conference on formal ontology in information systems, Saarbrücken, Germany, 2008.

Fernandez-Lopez, M. Gomez-Perez, A. and Juristo, N. "METHONTOLOGY: From Ontological Art Towards Ontological Engineering", paper presented at AAAI Symposium on Ontological Engineering, Stanford, USA, 1997.

Fox, M. S. Burbuceanu, M. and Gruninger, M. (1996) "An organization ontology for enterprise modelling: preliminary concepts for linking structure and behavior", *Computers in Industry,* Vol. 29, pp. 123–134

Gangemi, A. Catenacci, C. Ciaramita, M. and Lehmann, J. "Ontology evaluation and validation: An integrated formal model for the quality diagnostic task", Technical report, Laboratory of Applied Ontologies, Rome, Italy, 2005.

Gomez-Perez, A. "Ontology evaluation", In Handbook on Ontologies in Information Systems, edited by Steffen Staab and Rudi Studer, International Handbooks on Information Systems, chapter 13, 251–274, Berlin: Springer, 2004.

Gruber, T. R. (1995) "Towards principles for the design of ontologies usedfor knowledge sharing", *International Journal of Human-Computer Studies*, Vol.

Holsapple, C. W. Joshi, N. C. and Helmer, O. (1963) "An experimental application of the Delphi method to the use of experts", *Management Science*, Vol. 9, No. 3, pp. 458-467

Hsu, C. C. and Sandford, B. A. (2007) "The Delphi Technique: Making Sense Of Consensus, Practical Assessment", *Research & Evaluation*, Vol. 12, pp. 1-8.

Hulden, M. "A practical approach on creating a restricted ontology for crop wild relatives, Agricultural Information and Knowledge Management Papers", Project Report, 2007.

Lovrenčić, S. and Čubrilo, M. "Ontology Evaluation – Comprising Verification and Validation", paper presented at the 19th Central European Conference on Information and Intelligent Systems, Varaždin, Croatia, 2008.

Lozano-Tello, A. and Gomez-Pérez, A. (2004) "Ontometric: A method to choose the appropriate ontology", *Journal of Database Management*, Vol. 15, No. 2, pp. 1–18

Obrst, L. Ceusters, W. Mani, I. Ray S. and Smith, B. "The evaluation of ontologies", In Revolutionizing Knowledge Discovery in the Life Sciences, edited by J.O. Christopher Baker and Kei-Hoi Cheung, chapter 7, 139–158. Berlin: Springer, 2007.

Suárez -Figueroa, M. Gomez-Perez, A. and Fernandez-Lopez, M. "The NeOn Methodology for Ontology Engineering", In Ontology Engineering in a Networked World, 9-34, Berlin Heidelberg: Springer, 2012.

Sure, Y. and Studer, R. "On-To-Knowledge methodology", In On-To-Knowledge: Semantic Web enabled Knowledge Management, edited by J. Davies and J. Wiley and Sons, pp. 42-47, 2002.

Tartir, S. Arpinar, I. B. and Sheth, A. P. "Ontological Evaluation and Validation", In Theory and Applications of Ontology, edited by Springer Netherlands, 115-130, 2010.

Völker, J. Vrandeciˇ, C. D. Sure, Y. and Hotho, A. (2008) "AEON – An approach to the automatic evaluation of ontologies", *Journal of Applied Ontology*, Vol. 3, No. 1-2, pp. 41–62

Zemmouchi-Ghomari, L. and Ghomari, A. R. "Reference Ontology", paper presented at the International IEEE Conference on Signal-Image Technologies and Internet-Based System, Marrakech, Morocco, 2009.

CHAPTER SEVENTEEN

ONTOLOGY-DRIVEN MVC: A VARIANT OF MVC ARCHITECTURAL STYLE

AMRITA CHATURVEDI AND T.V. PRABHAKAR

Despite separation of concern provided by the MVC architectural style, there still remain some inter-component dependencies which constitute the liabilities of the style. In this paper we highlight the drawbacks of the MVC architectural style and propose an ontology driven solution to address these problems. In the proposed solution the ontology is used as an active run time component. The ontology is generic and can be used for any application based on MVC style. We demonstrate the usage of this style through an example implementation based on both the traditional MVC style and the ontology-driven MVC style. We evaluated the impact of ontology on the quality attributes of the MVC style. We observe that the proposed solution apart from addressing the drawbacks of MVC style also enhances maintainability and reusability of an information system based on the proposed variant style.

Introduction

MVC (Model-View-Controller) style is an architectural pattern first conceived by Trygve Mikkjel Heyerdahl Reenskaug in 1979 for Graphical User Interface (GUI) design. The main goal was "to support the user's mental model of the relevant information space and to enable the user to inspect and edit this information" (Reenskaug 2003). MVC provides clear separation of concerns by decomposing an application into three parts: the model, the view and the controller. The model contains application data and the logic for accessing and manipulating that data. It notifies the view when a state change occurs in the model. The view is responsible for rendering the state of the model. The controller is responsible for intercepting and translating user inputs into actions to be performed by the model or view.

A pattern solves a particular problem but its application has various side effects, which may raise new problems (Buschmann et al. 1996)). Some of these can be solved by other patterns. MVC separates data and core functionality from the user interface, thus providing adaptable user interfaces. But this design decision makes the view dependent on the model state. To synchronize the view and model, a change propagation mechanism needs to be established so that the dependent views are updated automatically whenever the state of the model changes. The observer pattern (Gamma et al. 1996) has been largely used to provide this mechanism where the model embodies the role of the subject while the dependent views play the role of observers. However, several drawbacks of observer pattern have also been recognized (Gamma et al. 1996, Fowler 2006a). Besides that, there are several other dependencies between the M, V and C components in the MVC style, which need to be handled, or else they constitute the liabilities of the style. All such drawbacks of the MVC style are discussed in the section named as 'MVC Architectural Style – Problems and Dependencies'.

Ontology is one of the most widely used semantic technologies and is formally defined as "an explicit specification of a shared conceptualization" (Gruber 1993). It can be used to model domain knowledge in the form of a set of concepts and the relationship between them and can be expressed in ontology languages like OWL, OIL, DAML etc.

We have used a generic ontology as a run time component in MVC architectural style to overcome its drawbacks. All the inter component dependencies in the MVC style are encoded in the ontology. So that the components (model, view and controller), instead of directly accessing each other, access the ontology to draw the information about the component they wish to access and then accordingly invoke the methods of the component based on the information provided by the ontology. This approach not only overcomes the drawbacks associated with the MVC style but also enhances maintainability and reusability of an information system based on the style.

The paper is structured as follows: section 2 covers the related research on variants of MVC style and how ontologies are used in this style. Section 3 highlights the problems and dependencies lurking in the MVC style. Section 4 describes the ontology driven MVC style – the proposed solution. Section 5 summarizes the conclusion and ultimately the references are enlisted.

Related Research

Tactical Variants of MVC Style

Flow Synchronization (Fowler 2006b) uses explicit updating of screens sharing domain data while Observer Synchronization (Fowler 2006b) uses implicit updating (event invocation) of screens. Presentation Model (PM) or Application Model (Fowler 2006b) and Model – View – ViewModel (MVVM) (Gossman 2005) extract the state and behavior of the View out into a Model class with the difference that ViewModel also provides a specialization of the Model that the View can use for data-binding. Both PM and MVVM enhance testability while MVVM further decouples Model and View. To enhance testability, the View is made thinner in Supervising Controller (Fowler 2006b) and Passive View (Fowler 2006b). Unlike Supervising Controller, in Passive View, Controller is responsible for updating the Views so no synchronization work is required to be done by the View. Model – View – Presenter (MVP) (Potel 1996) and Model – View – Adapter (MVA) (Simler 2009) further decouple the Model and View by placing Presenter and Adapter components respectively in between the Model and View. Unlike MVA, MVP also separates the UI from the presentation logic, thereby enhancing reusability and testability. Presentation – Abstraction – Control (PAC) (Buschmann et al. 1996) defines a structure for interactive software systems in the form of a hierarchy of cooperating agents. Every agent is responsible for a specific aspect of the application's functionality and consists of three components: Presentation, Abstraction, and Control. This subdivision separates the human-computer interaction aspects of the agent from its functional core and its communication with other agents. The Hierarchical – Model – View – Controller (Cai et al. 2000) pattern decomposes the client tier into a hierarchy of parent-child MVC layers. HMVC is similar to PAC with the difference that model and view are not totally decoupled in HMVC. openMVC (Barrett and Delany 2004) is a framework based on the MVC pattern which allows the Style Information, Layout and Validation Constraints components to be updated with no code change or recompile requirement. This variant reduces redundancy and facilitates faster maintenance.

MVC Variants with Additional Component

Jorg Stocklein et al. (Stocklein 2009) extend the MVC design pattern by an additional dimension "Environment" to capture elements and constraint from the real world. Although the real world sensor information can be handled as Controller events in the MVC model, but this can lead to complex and obscure models. This extension (MVCE) helps in iterative design and development of next generation user interfaces where each component can be refined independently. MVCD (Model-View Controller-Data) style (Zhang 2006) just abstracts the data source to form an independent data layer. The data layer transforms data received from the model layer into data in the same storage format with the appointed database and stores it to the database. Therefore, the MVCD model can be independent of database. The MVCD model inherits various kinds of advantages of MVC model and has better expansion and maintainability. Chen and Han (Chen and Han 2008) present an improved MVC style where the controller draws out the parameters from the user request objects, and then makes an instantiation or obtains a corresponding Servlet in memory according to the parameters to respond the request. The corresponding Servlet is called Action in the structure. The servlet inherits from abstract Action class and overrides the run method to respond to the particular request. This improvement enhances maintainability of the style.

Use of Ontology in MVC

Bolognini et al. (Bolognini et al. 2009) employ ontology based MVC style to implement a sophisticated, user-friendly, generic, customizable metadata editor. The proposed metadata editor uses different OWL ontologies to create custom forms for the input of instance values for an ontological metadata schema. The separation of the ontologies makes it possible to create metadata editors for any schema, and to customize the forms themselves to provide the best user experience in filling in the required metadata values. The main use of ontologies in this metadata editor is to generate user interfaces, which are generic and Model (metadata schema) independent. Gómez-Pérez and López-Cima (Gómez-Pérez and López-Cima 2004) present a rapid web application development framework to build semantic portals based on MVC style. In the proposed variant, the Model is composed of domain and cross-domain ontologies and the Controller has ontology based navigation model for controlling the user's navigations. The main difference between this work and ours is that the described navigation ontology is domain specific and cannot be reused, contrary to our MVC ontology which is domain independent. Also, the ontologies described are only helpful in navigating the user from one view to another or generating a composed view. Unlike this work, our variant also deals with model related user actions, model – view – controller dependencies and view notification mechanism. Paulheim (Paulheim 2009) presents a framework that uses ontologies for building user interfaces from independent, loosely coupled modules (plugins). The information describing a plugin is distributed across three ontologies namely the application ontology, domain ontology and plugin ontology. Plugins communicate with each other through event exchange mechanism. It provides an extension of MVC style where each plugin consists of at least four parts viz. Model, View, Controller and Plugin Ontology, which defines the components and interactions of the plugin. One reasoning component is also there responsible for processing the different ontologies and triggering the appropriate events. User action events generated by a plugin's user interface are read by the reasoner component. The reasoner queries the ontologies for components defining interactions triggered by this type of event, and notifies the respective plugins. It proposes a centrally mediated communication pattern between different components (plugins) which are themselves based on MVC architectural style. It basically employs ontologies for removing inter-component dependencies where each component is based on MVC style. Unlike this, our paper deals with intra component dependencies in an MVC style and attempts to remove its drawbacks by proposing ontology driven solution.

MVC Architectural Style: Problems and Dependencies

Problems with the MVC Style

The MVC architectural style was designed to balance two forces:

(1) It should be easy to modify the user interface
(2) The functional core of the software should not be affected by user interface modification

In an attempt to balance these forces, the core functionality, input and output were separated and encapsulated in three components namely model, controller and view respectively. But this design gives rise to another problem by making the view dependent on the model state. Whenever the model changes, all its dependents must be updated accordingly. This change propagation mechanism is largely implemented by employing the observer pattern where the model embodies the role of the subject while the views play the role of observers.

However, there are several drawbacks associated with observer pattern (adapted from (Gamma et al. 1995) and (Fowler 2006a)):

- **Implicit coupling between the view and the model:** Although the main intent of observer pattern is to minimize coupling but still there remains an implicit coupling between the model and view. The model has all the observing views registered with it, which need to be notified when a change occurs in the model.
- **Unexpected updates:** A small operation on model may lead to cascade of updates to observers and their dependents. Also since the dependency criteria aren't well defined or maintained, it may lead to spurious updates which may be hard to track down.
- **The push and pull model of updates:** In the push model, the model sends detailed information about the change to the views whether they want it or not. In the pull model, the model sends minimal notification and views ask for details explicitly thereafter. The push model increases coupling because the model must have some information about the views to send detailed information about the change. The pull model may be inefficient because the views must ascertain what has changed without the help of the model.
- **Dangling references:** When a view (observer) registers with the model, the model actually carries a reference to the view. If the view (observer) doesn't remove itself from the model when the screen is closed, it gives rise to zombie or dangling references and a memory leak.
- **Implicit event invocation:** When there is chain of event invocations, it becomes hard to follow and debug by looking at the code.

The Model, View and Controller Dependencies

Although MVC style separates business logic from presentation, yet there still remain some implicit dependencies which need to be handled:

(1) Views depend on each other and/or a parent child relationship exists between the views. There might be system requirement to update the parent views before their children or some particular views might have higher update priority. An example of view and sub – view relationship is shown in Figure 1. It shows a partial view hierarchy of General Demographic Survey Information

System. In Figure 1, the boxes denote the views and the arrows denote the sub – view relationship.

(2) The controller is responsible for converting the user action into system response. To do so, it directly calls the model and/or view methods to generate an appropriate response upon receiving the user input. This makes the controller dependent on the model and view interfaces, as a result a change in model and/or view interface tends to break the controller code.

(3) The model needs to have some knowledge of its observing views in order to notify them of changes. This implicitly couples the model and view.

Ontology-Driven MVC Architectural Style: The Proposed Solution

All the problems and dependencies discussed in the previous section can be solved by introducing ontology as a runtime component in the MVC style (and hence the name Ontology-Driven MVC). This section describes the generic Model – View – Controller Ontology and the static, dynamic and implementation aspects of the proposed style along with its consequences.

The Model – View – Controller Ontology

In order to model the MVC domain, we have created a generic ontology, which we call Model – View – Controller (MVC) Ontology. The MVC ontology comprises of MVC domain concepts like Model, WindowView, Controller, UserAction, UpdatePriority and source code concepts like Class, Constructor, Method, Query, VariableType. The MVC ontology maps the MVC domain concepts with the source code concepts and hence encodes the entire dynamic flow of the MVC style. Figure 2 shows a partial view of the MVC ontology concepts and Object Properties connecting two concepts. For any given user action (UserAction), the ontology returns the response method (Method) along with its class (Class) name. When a particular model (Model) information changes, the ontology returns the views (WindowView) displaying the information that has changed along with the updating methods (Method), classes (Class) and update priorities of the corresponding views.

Several MVC frameworks like Struts (McClanahan 2000) and Spring (Johnson 2003) use XML based routing configuration file to map user action to system response. But the MVC ontology in our proposed solution is employed not merely for configuration or metadata purposes but for modelling the MVC programming domain which provides several benefits over the XML based configuration file. The object properties along with their characteristics (Transitive, symmetric property) and domain – range restrictions define the MVC space concept relationships. The cardinality restrictions on the properties aid the developer in verifying the consistency of the resulting instantiated ontology. Different atomic views 'compose' (transitive object property) the compound and complex views which can be refreshed if they display the updated information rather than refreshing the entire view. Moreover, if the response to a particular user action requires that a view be displayed and if there is an error displaying the view, then any other view rendering similar information can be displayed. Moreover, the ontology also enables the application to perform priority based updating of the views rendering the model information that has changed during the runtime. Figure 3 shows a partial view of the MVC ontology concepts along with is-a relationship and instances. The portion of the image above the red dotted line shows the intension and that below shows the extension part of the MVC ontology. In the following text MVC ontology is simply referred to as 'ontology'.

Figure 6: Partial Hierarchy Showing Views and Sub-views of General Demographics Survey Information System

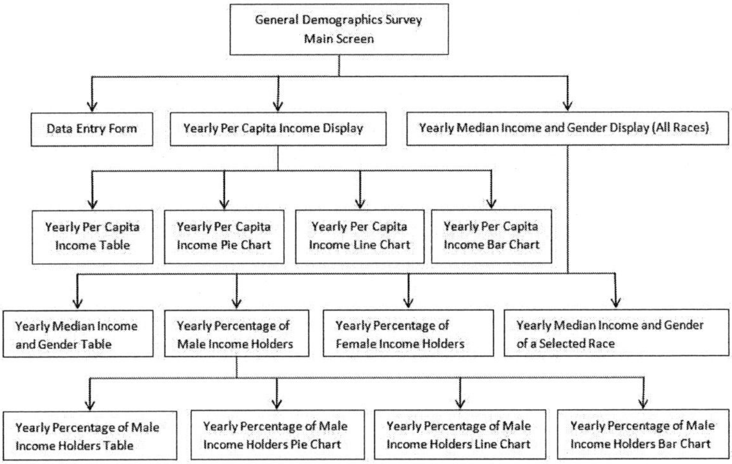

Figure 2: Partial View of MVC Ontology Generated by the Ontoviz Plugin of Protégé 3.4

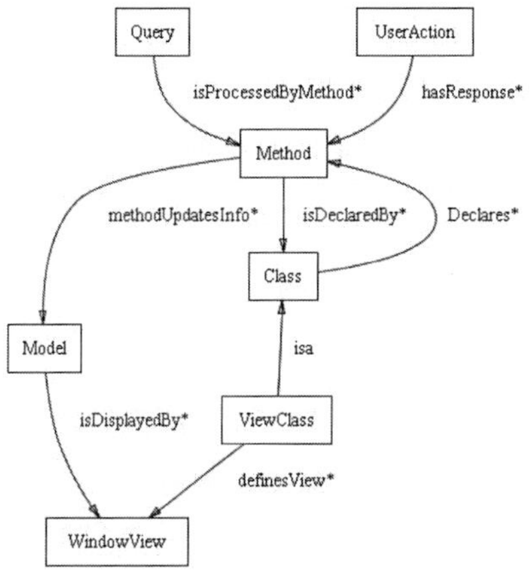

Structure

Figure 4 shows the difference between the structure of the traditional MVC style and the ontology-driven MVC style. In Figure 4, the numbers along the edges denote the sequence of communication. In the traditional MVC style (Figure 4 (a)):

(1) The controller receives the user action/input from the view.
(2) The controller calls the model and/or view methods to generate the desired system response.
(3) The model notifies the view of change in model state and also passes the changed state.
(4) The view may call model method to query the model state.

The proposed ontology-driven MVC style consists of four components viz. Model, View, Controller and Ontology with each one of them having their own APIs. The entire communication dependencies between the components are encoded in the ontology, which is domain independent. The sequence of communication between the components occurs as follows (Figure 4 (b)):

(1) The controller receives the user action/input from the view.
(2) The controller consults the ontology to find the system response for that particular input.

(3) The ontology sends the method information which should be invoked to achieve the response.
(4) The controller invokes the appropriate methods of the model or view to achieve the desired output.
(5) If in this process the model state is changed, then the model's NotifyAll method consults the ontology to find out the views rendering the information that has changed.
(6) The ontology sends the view information of those views which need to be updated.
(7) The NotifyAll method then accordingly, passes the changed information and invokes the updating methods of the views returned by the ontology in the order of their update priorities (as supplied by the ontology component).

Thus the ontology plays the pivotal role in the entire working of the style from receiving input to delivering output.

Figure 3: Partial View of MVC Ontology Showing the Concepts as well as is-a Relationship and Instances

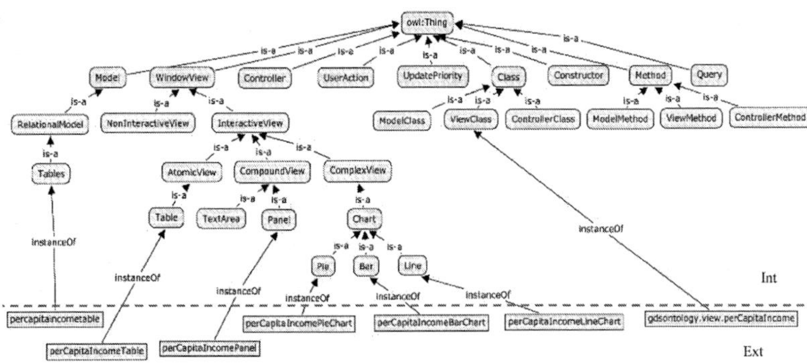

Figure 4: The Structure of (a) Traditional MVC Style (b) Ontology-Driven MVC Architectural Style

(a)

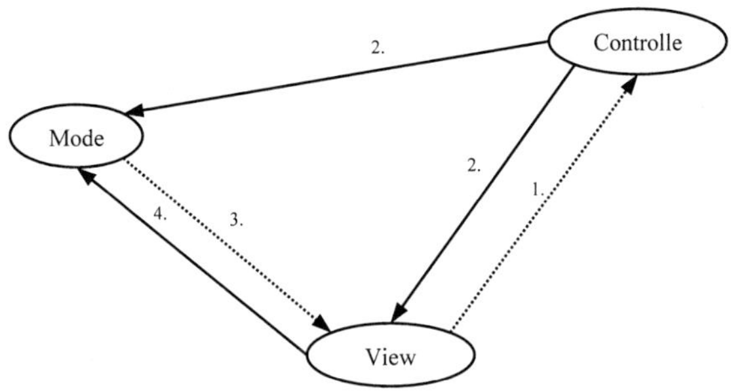

(b)

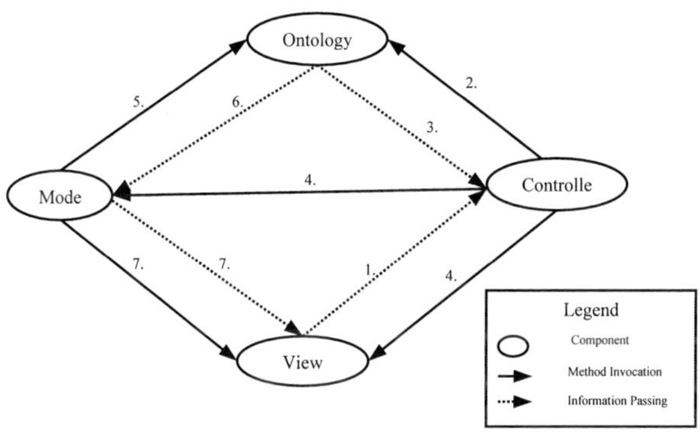

Dynamics

We have taken three common scenarios to depict the dynamic behavior of the proposed architectural style.

Scenario I

This shows how the controller gets the data from the model desired by the user (Figure 5).

- The controller handles the request of the user to get model data.
- The controller retrieves the query from the ontology which should be executed to get model data.

- The controller retrieves the model method from the ontology which should be executed to get the data.
- The controller retrieves the model class defining the method from the ontology.
- The controller then, invokes the model method whose detail was derived from the ontology.
- The model method after completing execution, returns the data to the controller.
- The controller then, returns from its getModelData method.

Figure 5: Sequence Diagram Showing Interaction between the Components to Get Desired Data from the Model

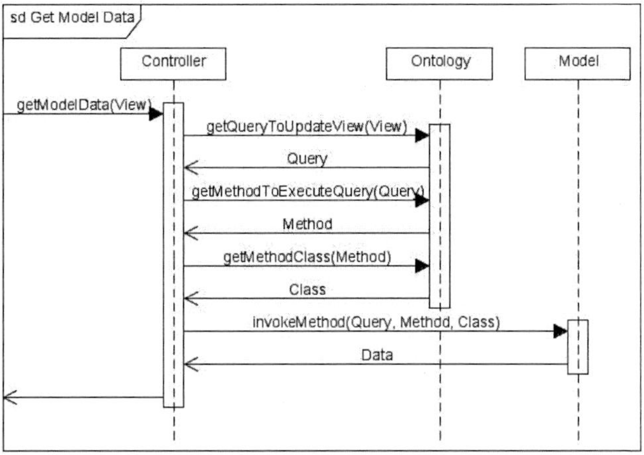

Scenario II

This shows how the controller updates the model data (Figure 6).

- The controller handles the user request to set model data.
- The controller retrieves from ontology, the model property updated by the view through which the request is sent.
- The controller retrieves the model method from ontology which updates the model property.
- The controller retrieves the model class defining the method from the ontology.
- The controller then, invokes the model method whose detail was retrieved from ontology.

- The model method, after updating the model property, calls the NotifyAll method of the model which is responsible for notifying all the views rendering the model property that has changed/updated.
- The NotifyAll model method retrieves from the ontology the views rendering the model property which is updated.
- The NotifyAll model method calls it's sub-routine to arrange the views according to their update priority.
- For each view, the updating method is retrieved from the ontology and is invoked by the NotifyAll model method.

Figure 6: Sequence Diagram Showing Interaction between the Components to Update the Model

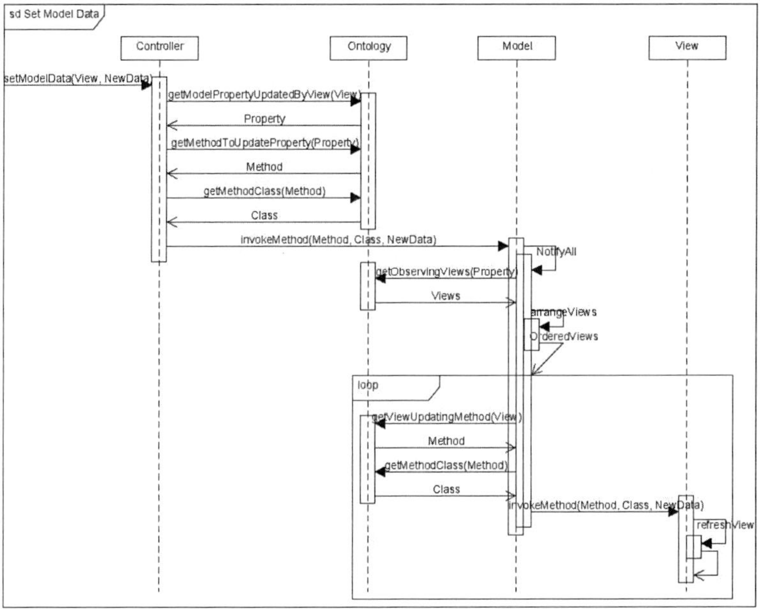

Scenario III

This shows how the controller displays views desired by the user (Figure 7).

- The controller handles the user request to display a view.
- The controller retrieves the response method, method type and method class from the ontology.
- The controller then, invokes the view method whose detail was retrieved from the ontology.

Figure 7: Sequence Diagram Showing Interaction between the Components to Display the View Desired by the User

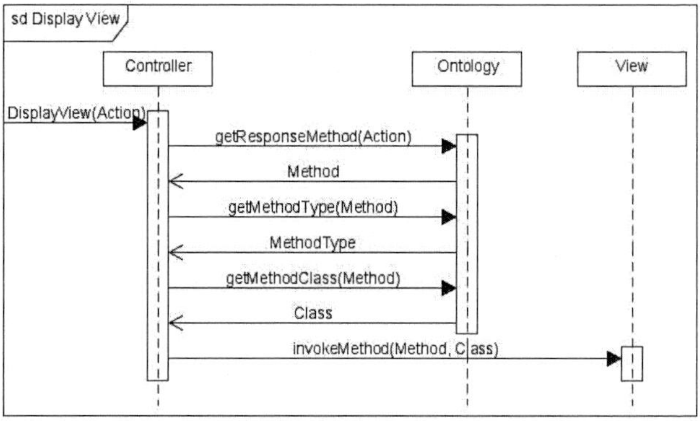

Implementation Aspects

The proposed architectural style gives rise to a new programming technique where the main principle is to encode the inter component dependencies in ontology. The following steps should be followed to develop an application based on the proposed architectural style:

(1) Design and develop the model: Encapsulate the data and core functionality of the application in the model component. Define functions for accessing and manipulating the data and expose the desired functions of the model to the controller.

(2) Design and develop the views: Design the appearance of main views as well as their children and dependent views. Dedicate a view class to initialize all the views rendering model data.

Implement the draw and update methods to display and update the views respectively. The update method is called by the model method performing notification on model state change.

(3) Instantiate the MVC ontology based on the application: The MVC ontology used for the proposed architectural style is generic and can be used for any application based on MVC architectural style. The ontology needs to be instantiated based on the different views, models and their methods used. A partial view of concepts and their relationships in the MVC ontology is shown in Figure 2 and instances of some concepts are shown in Figure 3.

(4) Design and implement a generic and reusable controller or extend and/or reuse the controller: The controller used in the proposed architectural style is completely generic (does not depend on the application) and is reusable. The controller can be implemented in any desired programming language following the principles laid down in the proposed solution. The resulting controller is reusable and has methods defined like setModelData, getModelData, displayView which are ready to use. Moreover, the controller methods do not depend either on the model or view interfaces. They rather derive their desired information from the ontology and hence enable the controller to be reusable with any set of models or views.

(5) Reuse the NotifyAll method of the model: The NotifyAll model method initiates the change propagation mechanism. When the model state changes, the NotifyAll method retrieves the dependent views from the MVC ontology and invokes their update methods. It derives the required information from the ontology which makes it application independent and reusable.

We have identified the software quality attributes affected by the proposed solution through a General Demographics Survey (GDS) information system based on the proposed architectural style. The information system is coded in java programming language and jena2.5.7 (Jena) is used as the ontology API. Our MVC ontology is in OWL – DL (W3C OWL Working Group 2004) and is developed using the protégé 3.4 ontology editor (protégé ontology editor). We also developed the same information system based on the traditional MVC style and then analyzed and compared the quality attributes exhibited by both the styles. Our findings are illustrated in the next subsection.

Consequences

To analyze the quality attributes affected by the proposed solution, we use the Dromey's software quality model (Dromey 1995) to map the characteristics exhibited by the information system based on the traditional MVC style as well as on the proposed style to high level quality attributes. Among the several software quality models like ISO/IEC standard (ISO/IEC 2010), IEEE standard (IEEE Computer Society R2009), we have chosen the Dromey's software quality model because it is the only model which focuses on software code or implementation. The model advocates that software does not manifest quality attributes. It, rather exhibits certain characteristics which can be mapped to high level quality attributes. Since we have software code at hand, we can easily apply the Dromey's model to obtain the high level quality attributes exhibited by softwares based on two different architectural styles.

Benefits

Since the main identified drawbacks with the MVC style deal with inter – modular issues and the proposed solution targets to solve the identified problems so among the four basic quality carrying properties identified by Dromey (Dromey 1995), the properties affected by the proposed solution are modularity properties (high level, inter – modular design issues). Among the six modularity properties identified, we find that the proposed solution enhances three of them which are parameterized, loosely coupled and abstract.

(1) Parameterized: "A module is said to be parameterized if it contains as parameters all and only the necessary and sufficient inputs and outputs to characterize a particular well-defined function/procedure" (Dromey 1995). Unlike the traditional MVC style, our proposed solution offers a completely parameterized controller and NotifyAll part of the model. In the controller, the parameters define the user action and the system response while in the NotifyAll the parameters define the entire notification mechanism when some model property changes along with the changed model property. The high level quality attributes affected by this property are maintainability and reusability (Dromey 1995).

(2) Loosely Coupled: We use the JArchitect tool (JArchitect) to analyze the java code of our information system based on the MVC style and on the Ontology-Driven MVC style separately.

Figures 8 (a) and 8 (b) show the dependency cycle graph between the components of the GDS Information System based on the traditional MVC style and on the Ontology-Driven MVC style respectively. Single arrow edge from A to B indicates that A is using B and double arrow edge indicates that A and B are mutually dependent. The thicknesses of edges are proportional to the strength of coupling in terms of the number of members involved. Figure 8 (a) shows that there exists a strong coupling between the gds.model, gds.view and gds.controller components and they are mutually dependent. Figure 8 (b) shows only a single arrow from gdsontology.view to gdsontology.controller which shows that view accesses the generic controller methods to hand over the user action. Apart from this, no other dependency exists between the model, view and controller components. Figures 9 (a) and 9 (b) show the dependency matrix between the components of the GDS Information System based on the traditional MVC style and on the Ontology-Driven MVC style respectively. A black matrix cell denotes that the package in row and the package in column are both directly using each other. The number 2 in black matrix cell denotes that A is using B and B is using A so A and B are involved in a dependency cycle of length 2. A blue matrix cell denotes that the package in column is using the package in row and a green matrix cell denotes that the package in column is used by the package in row. There are 6 black matrix cells in Figure 9 (a) which means that all the three model, view and controller components are engaged in a dependency cycle of length 2. Whereas, there are no black matrix cells in Figure 9 (b) so there is no mutual dependence between the model, view and controller components. The high level quality attributes affected by this property are maintainability, reusability and reliability (Dromey 1995).

(3) Abstract: "An object/module is sufficiently abstract if there is no obvious, useful higher level concept that encompasses the structural form" (Dromey 1995). The controller and NotifyAll part of the model are sufficiently abstract as very high level generic methods are used for handling input, delivering output and notifying the observers. The high level quality attributes affected by this property are maintainability and reusability (Dromey 1995).

Figure 8: Dependency Cycle Graph between the Components of the GDS Information System based on (a) The Traditional MVC Style (b) The Ontology-Driven MVC Style

(a)

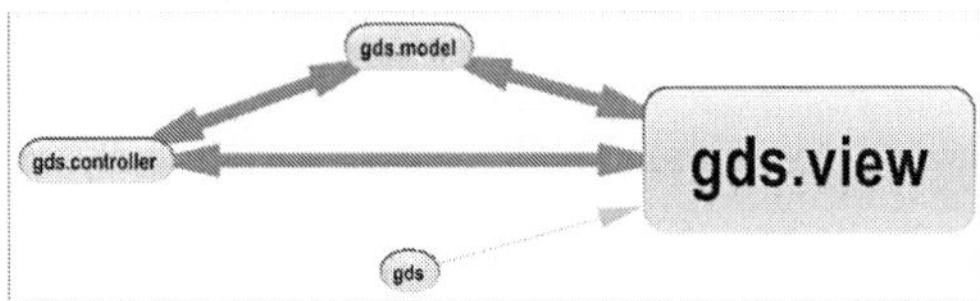

(b)

Figure 9: Dependency Matrix between the Components of the GDS Information System based on (a) The Traditional MVC Style (b) The Ontology-Driven MVC Style

(a) (b)

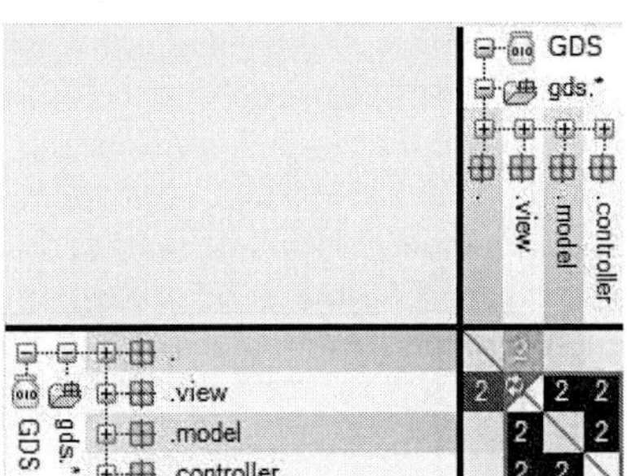

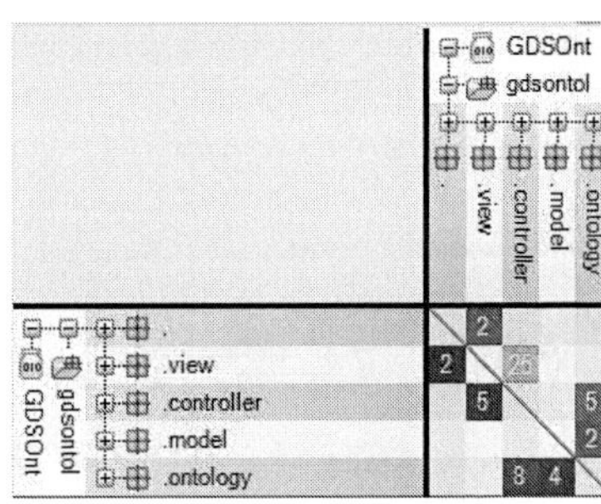

Apart from the added enhancement of the above mentioned quality attributes, the proposed solution also overcomes the drawbacks associated with the MVC style (discussed in section 3). This summary is presented in table 1.

Table 1: Table showing how the proposed solution overcomes the drawbacks of MVC style

MVC style Drawbacks	Proposed Solution
Implicit coupling between the view and the model	No coupling exists between view and model as shown in Figures 8 (b) and 9 (b)
Unexpected updates	Since the dependency criteria of observers are well defined and maintained in the ontology, there is no chance of spurious updates.
The push and pull model of updates	Since the entire notification information is provided by the ontology, there is no need of push or pull model of updates
Dangling references	Since the observers don't register themselves with the model, there is no risk of dangling references.
Implicit event invocation	Since there is no coupling between the model, view and controller (Figure 8 (b) and 9 (b)), there is no chance of complicated event invocations.

Liabilities

Although the proposed solution enhances certain quality attributes, but it also causes an indirection in the flow of control and uses reflection technique and hence negatively impacts performance. As shown in Figure 10, memory consumption is more in case of ontology-driven MVC style because of the processing of the ontology involved.

Although, to develop an information system based on the proposed style, the developer has to instantiate the ontology once but the effort is recovered over multiple implementations in the maintenance phase.

Figure 10: Runtime Memory Heap Graph of GDS Information System based on (a) The Traditional MVC Style (b) The Ontology-Driven MVC Style

(a)

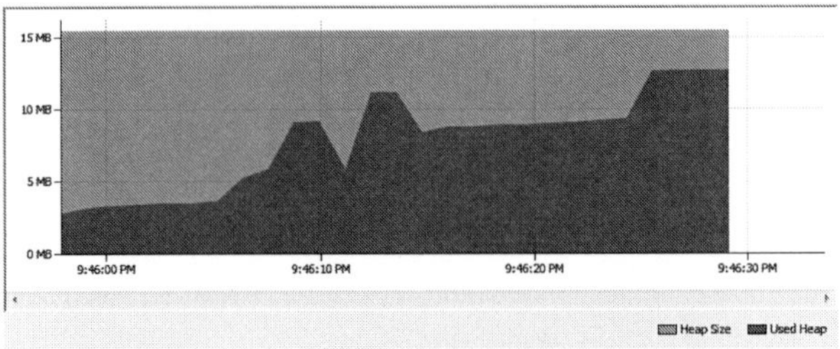

(b)

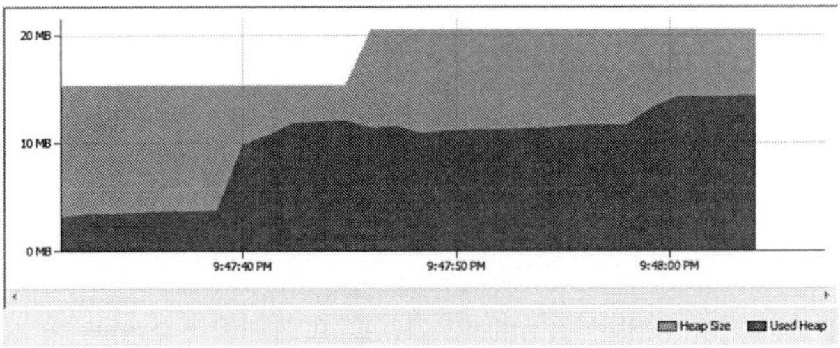

Conclusion

In this paper, we have highlighted the problems of the MVC style and have proposed an ontology driven solution to address the identified problems thus defining a new variant of MVC architectural style: Ontology-Driven MVC architectural style. In the proposed solution the ontology is used as an active run time component. The ontology is generic and can be used for any application based on MVC style. All the inter component dependencies in MVC style are encoded in the MVC ontology so that the components instead of directly accessing each other, draw the communication details from the ontology and subsequently invoke appropriate methods. We have demonstrated the utility of the proposed style through an example implementation based on the proposed style. We found that the proposed solution addresses all the drawbacks of MVC style and also enhances software quality attributes like maintainability and reusability. The implementation aspects reveal that the proposed solution gives rise to a new programming technique where the ontology drives the entire communication sequence, breaking the communication dependencies between the components. This enhances the maintainability and reusability of the components but negatively impacts the overall performance because of the indirection in the flow of control and use of reflection technique. The ontology-driven MVC style can be used in the following contexts:

(1) When the model and/or the view are expected to change in near future: For example, the model of the GDS information system may have to be changed by recording the educational attainment of the population in addition to their mean income. In this scenario apart from the model changes, the view also requires minor changes. There is another scenario where the view requires major changes but the model remains unchanged. This is the case when additional presentations of the same data have to be made like adding 3D pie charts, animated charts, histograms etc. There is another scenario where both the model and view need to be changed. For example the GDS information system is to be converted into an election information system for recording and displaying the election data. In this worst-case scenario, both the view and model are implemented and connected with the reusable controller and notification mechanism via the MVC ontology.

(2) When the view is rich: This is the case when there are several widgets on the same screen displaying different parts of the model and handling several user actions. For example in a medical information system, one text area may be displaying symptoms of a disease and another text area displays the recommended medicines for the same disease. When the user clicks a medicine, a third text area displays the effects and side effects of the medicine. Such rich widgets and user actions can be better developed and maintained by being declared in the MVC ontology rather than defined imperatively in scattered code.

(3) When runtime changes in the view can be made: This scenario arises when the user is allowed to change the widgets at runtime. The response to all user actions is encoded in the MVC ontology and hence this can be flexibly done.

(4) When the user interface (UI) designer wants to generate and fix various different types of user interfaces: Since the system response to various user actions is encoded in the MVC ontology, the designer can flexibly change the structure of UI and test run the various UI designs

It should be possible to reduce the overhead imposed by the ontology by efficient implementations of this component using main-memory based RDF or RDBMS components.

Acknowledgments

The authors would like to thank Indian Institute of Technology, Kanpur, Uttar Pradesh, India for providing sufficient infrastructure and resources which aided in the successful completion of this work. We would also like to thank Dr. Philippe Dugerdil, Professor of Software Engineering, Department of Information Systems, Geneva School of Business Administration, University of Applied Sciences, Geneva, Switzerland for his valuable comments and feedback.

References

Barrett R.; Delany S. J. (2004) "openMVC: A non-proprietary component-based framework for web applications" in *WWW 13th International World Wide Web Conference*, pp. 464-465.

Bolognini V.; Iorio A. D.; Duca S.; Musetti A.; Peroni S.; Vitali F. (2009) "White B.: Exploiting ontologies to deploy user-friendly and customized metadata editors" in *Proceedings of the IADIS Internet/WWW conference*, Rome, Italy.

Buschmann F.; Meunier R.; Rohnert H.; Sommerlad P.; Stal M. (1996) *Pattern-Oriented Software Architecture A System of Patterns*, volume 1. Wiley.

Cai J.; Kapila R.; Pal G. (2000) *HMVC: The layered pattern for developing strong client tiers*, http://www.javaworld.com/javaworld/jw-07-2000/jw-0721-hmvc.html?page=2, Accessed on 21 April 2012.

Chen C.; Han H. (2008) "The research on modern distance education system based on improved mvc pattern" in *International Conference on Computer Science and Software Engineering*, Vol. 5, pp. 462-465.

Dromey R. G. (1995) "A model for software product quality", *IEEE Transactions on Software Engineering*, Vol. 2, pp. 146-163.

Fowler M. (2006a) *Organizing presentation logic: observer gotchas*, http://martinfowler.com/eaaDev/OrganizingPresentations.html#observer-gotchas, Accessed on 20 March 2012.

Fowler M. (2006b) *GUI architectures*, http://martinfowler.com/eaaDev/uiArchs.html, Accessed on 20 March 2012.

Gamma E.; Helm R.; Johnson R.; Vlissides J. (1995) *Design Patterns Elements of Reusable Object-Oriented Software*, Addison-Wesley.

Gómez-Pérez A.; López-Cima A. (2004) "Advanced design of semantic web portal: ODESeW2" in *Poster Proceedings of 3rd International Semantic Web Conference (ISWC2004)*, Hiroshima, Japan.

Gossman J. (2005) *Introduction to Model/View/ViewModel pattern for building WPF apps*, http://blogs.msdn.com/b/johngossman/archive/2005/10/08/478683.aspx. Accessed on 21 April 2012.

Gruber T. (1993) "A translation approach to portable ontology specification", *Knowledge Acquisition*, Vol. 5, No. 2, pp. 199-220.

Guarino N. (1998) "Formal Ontology and Information Systems" in *Proceedings of Formal Ontology in Information Systems (FOIS '98)*, IOS Press, Trento, Italy.

IEEE Computer Society (R2009) *IEEE Standard for a Software Quality Metrics Methodology*, IEEE Std 1061-1998.

ISO/IEC (2010) *Systems and software engineering – Systems and software Quality Requirements and Evaluation (SQuaRE) – System and software quality Models*, ISO/IEC FDIS 25010 International Standard, http://www.iso.org/iso/iso_catalogue/catalogue_tc/catalogue_detail.htm?csnumber=35733, Accessed on 15 August 2012.

JArchitect: *A tool to evaluate java code base*, http://www.javadepend.com/. Accessed on 10 April 2012.

Jena: *A Java Framework for building semantic web applications*, http://sourceforge.net/projects/jena/files/Jena/Jena%202.5.7/, Accessed on 5 March, 2012.

Johnson R. (2003) *Spring framework, MVC based web application framework*, http://www.springsource.org/, Accessed on 25 April 2012.

McClanahan C. (2000): *Apache struts, MVC based web application framework*, http://struts.apache.org/2.x/index.html, Accessed on 25 April 2012.

Paulheim H. (2009) "Ontology-based modularization of user interfaces" in *Proceedings of the 1st ACM SIGCHI Symposium on Engineering Interactive Computing Systems*, pp. 23-28.

Potel M. (1996) *MVP: Model-View-Presenter The Taligent programming model for c++ and java*, http://www.wildcrest.com/Potel/Portfolio/mvp.pdf, Accessed on 21 April 2012.

Protégé ontology editor, http://protege.stanford.edu/, Accessed on 5 March 2012.

Reenskaug T. (2003) *The Model-View-Controller (MVC) its past and present*, http://hcim.ifi.uio.no/~trygver/2003/javazone-jaoo/MVC_pattern.pdf, Accessed on 2 March 2012.

Simler K. (2009) *Model – View – Adapter*, http://blog.palantir.com/2009/04/20/model-view-adapter/, Accessed on 21 April 2012.

Stocklein J.; Geiger C.; Paelke V.; Pogscheba P. (2009) "Poster: MVCE – a design pattern to guide the development of next generation user interfaces" in *IEEE Symposium on 3D User Interfaces*, pp. 153-154.

W3C OWL Working Group (2004) *Web ontology language overview*, http://www.w3.org/TR/owl-features/, Accessed on 5 March 2012.

Zhang L.; Ma F.; Zhong C; Zhang L.; Wang Y. (2006) "A MVCD model and its application in university project management" in *The Sixth World Congress on Intelligent Control and Automation (WCICA)*, Vol. 2, pp. 7008-7012.

CHAPTER EIGHTEEN

THE ROLE OF INFORMATION TECHNOLOGY IN E-LEARNING SYSTEMS SUCCESS[1]

SEAN EOM, NICHOLAS J. ASHILL, BEN ARBAUGH AND JAMES L. STAPLETON

Introduction

The global market for e-learning[2] reached US$27.1 billion in 2009 and The demand is growing by a five-year compound annual growth rate (CAGR) of 12.8% and revenues will reach $49.6 billion by 2014. But in spite of such potential promise for and increasing research support for the effectiveness of e-learning, support for delivering education through this medium is far from unanimous . Specific to business and management education, one of the potentially confounding factors is the extent to which characteristics of the delivery technology enhance opportunities for learning. To date, results of studies examining the effects of technological characteristics on course outcomes in online have been mixed. Generally, this literature suggests that although technological characteristics predict students' willingness to use and satisfaction with virtual learning environments as an educational medium, other factors such as learner

---

[1] First published in *Human Systems Management*, 2012, Vol. 31, Iss. 3, pp. 147-163.

[2] E-learning a.k.a., distance learning, or online learning, is defined as learning delivered via the Internet and Web technologies. E-learning is extending learning opportunities from a fixed temporal and geographical classroom to virtual classrooms available anywhere and anytime. Between two modes (e-learning and face-to-face learning) of instruction, there are many types of variation: blended and webinar courses. Blended courses refer to the one that combines the class meeting and online meeting. Another type of distance learning courses is webinar which requires synchronous class meetings be held online through audio/video technologies.

motivation, cultural characteristics, and/or peer support tended to be stronger predictors. Further, these studies have yet to produce definitive, consistent results regarding the relationship between technological characteristics and student learning.

However, it is possible that this lack of relationship between technological characteristics and course outcomes may be due to how technology is conceptualized rather than how it is deployed. In online business education, the dominant approach used to conceptualize technology has been the Technology Acceptance Model (TAM) . Although a highly influential model that has undergone further refinement , its original intent was to predict voluntary adoption of new technologies. Therefore, researchers seeking to use the model to explain outcomes in online courses other than attitudes toward the technology and usage are making a conceptual leap. Conversely, other MIS scholars have developed conceptual and theoretical frameworks of online learning effectiveness. However, these conceptual frameworks generally have gone untested . Therefore, it is possible that other conceptualizations of technology may yield stronger explanations for relationships between technological characteristics in online business education.

One model that may better explain the role of technology in online teaching and learning effectiveness is DeLone and McLean's (DM) model of information system (IS) success. The DeLone and McLean model consists of six major categories of success measures, which are interrelated and interdependent. These factors are system quality, information quality, use, user satisfaction, individual impact, and organizational impact. The DeLone and McLean model states that the system use is a mediating variable that is functionally determined by informational quality and system quality. The system quality is concerned with measures of the information processing system itself in evaluating the contribution of IS to the organization. These measures are multidimensional, including response time, systems accessibility, system reliability, systems flexibility, systems usefulness, and ease of learning, etc. Because the DM model operationalizes technological characteristics beyond the dimensions of usefulness and ease of use, can be applied more readily to environments where technology use is not voluntary, and considers outcomes in addition to technology adoption, it may be a more effective approach for conceptualizing technology in more comprehensive models of online learning effectiveness.

The purposes of this paper are twofold. First, this paper empirically tests the validity of IS success model of DeLone and McLean in a university online education context. The IS success model of DeLone and

McLean is one of the widely tested information systems models, if not the most widely tested model. Second, we discuss the broad implications of the results of our empirical assessment of the validity of the IS success model. The driver of e-learning systems is not e-leaning system's hardware and software itself but the users (instructors and students) and their learning and teaching strategies. The remainder of the paper is organized by five major sections. The validation of the IS success model in information systems context section reviews three studies that empirically test the validity of the DeLone and McLean model. This is followed by a section that presents conceptual framework and hypotheses derived from this model. We then present a section that deals with research methodology, including the description of survey instruments and partial least square (PLS) method of analysis. After reporting the results of measurement (outer) model and structural (inner) model in the paper's fourth section, our final section discusses the relation of our findings to prior research on online business education and its implications for future research.

Validation of the Delone and Mclean Model in an Information Systems Context

Ever since the field of management information systems (MIS) emerged as an academic discipline in 1960s, one of the pivotal issues to establish MIS as a coherent research area was the question of the dependent variable. As a system, it is necessary to define the output of the system and its measurement, and the process of transforming data to information (output). What are critical success factors of MIS success? To answer this question, DeLone and McLean reviewed 180 empirical studies and formulated a more integrated model of information systems (IS) success[3]. Their IS success model identified six constructs that are interrelated and interdependent – system quality, information quality, use, user satisfaction, individual impact, and organizational impact (Figure 1). The DeLone and McLean model consists of two antecedents constructs (system quality and information quality) and two intervening constructs (system use and user satisfaction). Use and user satisfaction are affecting each other reciprocally. These two intervening constructs are direct antecedents of individual impact.

---

[3] The term "success" is defined as accomplishment of desired outcomes. In the DeLone and McLean model, the desired outcomes are individual and organizational impacts.

Figure 1: The DeLone and McLean model

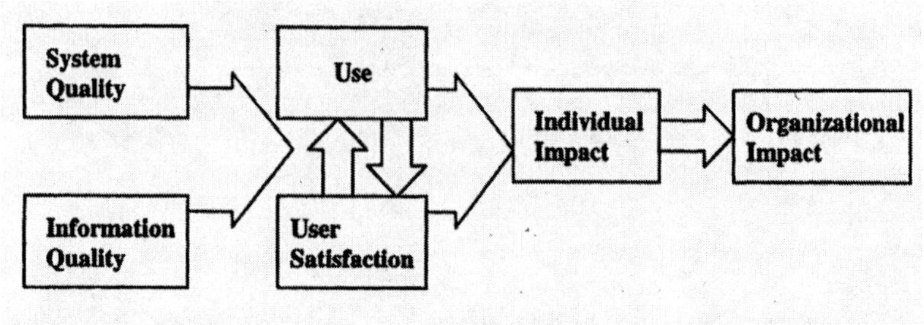

Reprinted by permission. Copyright[1992] INFORMS.

Although the DeLone and McLean model was proposed without empirical testing, three studies have sought to empirically test its validity. Rai and others tested the model in a quasi-voluntary IS use context using data collected from the users of an integrated student information systems at a mid-western university in the US. Both system quality and information quality positively affected systems use and user satisfaction, and each, in turn, contributed to information systems success. Rai et al. concluded that the DeLone and McLean model has explanatory power, and therefore, the model has merit for explaining IS success.

Livari conducted the second major study to test the DeLone and McLean model using a field study of a mandatory information system implemented in a municipal organization in Oulu City Council, Finland. The structural equation model however, partially supported the DeLone and McLean model as a predictive model of information systems success. Three important paths were not statistically significant; these were the relationships between system quality-use, information quality-use and between use and individual impact (hypotheses H_1, H_2, and H_7 in Figure 2). Nevertheless Livari concluded that "overall, the results support the reasonableness of the DeLone-McLean model as a predictive model" but acknowledged the need for further empirical testing in the case of voluntary systems.

Recently, the DeLone and McLean model was tested in the e-learning context using a confirmatory factor analysis (CFA) module of the LISREL software . All hypotheses presented in their research model were accepted with a significance level of 0.01. Based on the results of the structural equation model that showed that both system quality and information quality were significant predictors of user satisfaction and system use, they concluded that the DeLone and McLean model can be an effective initial model for evaluating an e-learning systems success. Freeze et al. argue that the basic assumption of voluntary use by the user is incorrect in the context of university e-learning courses where usage of the system is required to complete the coursework. However, we assert that the model tested by Freeze et al. is mis-specified because no specified relationships between system use and user satisfaction are considered. Even when usage of the system is required, variability in the quality and intensity of this use is still likely to have a significant impact on the realization of the system benefits. Furthermore, no system is totally mandatory. Consequently, these two missing links in the Freeze et al. study (the links from system use to user satisfaction and the reciprocal) may have resulted in misleading results when examining the impact of system use and user satisfaction on information systems success.

Conceptual Framework and Hypotheses

When testing the DeLone and McLean model of IS success in an e-learning context, an initial and important decision was choosing one of the two models (original vs. extended model). We decided to choose the original model of IS success due to the following reasons. First, the majority of e-learning systems are not home grown systems (e.g., blackboard, moodle). Second, e-learning students rarely interact with e-learning system development staff. Occasionally, some students experience problems with their own computers such as virus infection, booting failures, and other symptoms of various malfunctions. Third, tangibility and reliability of systems are included in the measurement of system quality.

The core of the DeLone and McLean model is that system use and user satisfaction are direct antecedents of individual impacts (Figure 2). DeLone and McLean state:

> SYSTEM QUALITY and INFORMATION QUALITY singularly and jointly affect both USE and USER SATISFACTION. Additionally, the amount of USE can affect the degree of USER SATISFACTION – positively and negatively – as well as the reverse being true. USE AND USER SATISFACTION are direct antecedents of INDIVISUAL IMPACT; and, lastly this impact on individual performance should eventually have some ORGANIZATIONAL IMPACT.

Figure 2: E-learning system success model to be tested

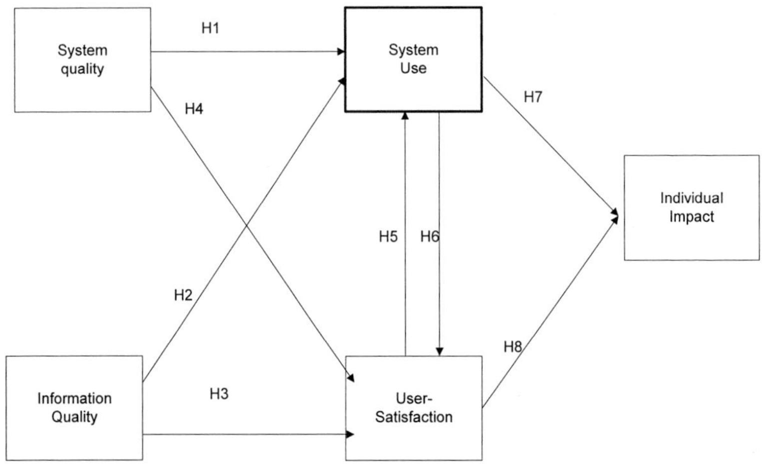

System quality

The IS success model and the e-learning success model posit that the success of IS and e-learning systems is dependent on the intervening variables (user satisfaction and system use), which are in turn dependent on the quality of information, system, and service.

System quality and system use

System quality is the measure of distinctive characteristics of information/e-learning systems, including convenience (ease of use) of access, response time, ease of learning, reliability, and perceived usefulness. Causal relationship between system quality and system use can be understood by TAM. The TAM theorizes that system use is determined by system quality (perceived usefulness and perceived ease of use). Perceived usefulness and ease of use determine an individual's intention to use a system, which in turn, determines actual system use. The theory posits that the four key constructs directly determine usage intention and behavior .

Empirical tests of the positive relationship between system quality and system use are documented in the context of mandatory IS use and in the quasi-volitional IS use. Therefore, we hypothesize:

H_1. There will be a positive relationship between system quality and e-learning systems use.

Systems quality and user satisfaction

Although DeLone and McLean proposed their model without empirical testing of temporal and causal relationships among the 6 constructs as a whole, it is built on the synthesis of previous empirical research that investigated relationships between all 6 constructs. Numerous studies support a positive relationship between system quality and user satisfaction in the case of voluntary systems such as decision support systems, mandatory systems , and quasi-volitional systems. In an e-learning systems context, Arbaugh found that perceived usefulness and ease of use of Blackboard significantly predicted student satisfaction with the Internet as an educational delivery medium. In light of existing research we hypothesize:

> H_2. There will be a positive relationship between e-learning system quality and student satisfaction.

Information quality

Information quality is the measure of distinctive characteristics of the quality of information produced by e-learning systems. Information quality consists of a range of attributes including accuracy, timeliness, reliability, relevance, format, perceived usefulness, and understandability .

Information quality and use

Unlike user satisfaction, empirical testing on the relationships between use and the quality of information and information systems has produced mixed results. Some studies have found that when the users are provided with high quality information in a voluntary systems context, they use information systems more frequently in a voluntary systems context , and in quasi-volitional systems . On the other hand, in an e-learning systems mandatory context, Livari failed to find such a causal relationship between the two. Since we are attempting to test the validity of the DeLone and McLean model, we used the original hypothesis of the DeLone and McLean model. Thus, in the current study we hypothesize:

> H_3. There will be a positive relationship between information quality and systems use.

Information quality and user satisfaction

DeLone and McLean define user satisfaction as "recipient response to the use of the output of an information systems." It is the response of users to indicate the level of confident acceptance of information as satisfactory, dependable, useful, true, and so on. It can be the difference between information needed and information received . Previous research unequivocally supports a positive relationship between information quality and user satisfaction in a mandatory system , and in a quasi-volitional system. Therefore, we hypothesize:

H_4. There will be a positive relationship between information quality and student satisfaction.

Use and user satisfaction

Use includes use of both the information system and output of information systems (information). Empirical measures include frequency of use, regularity of use, and actual amount of use. Both user satisfaction and use have been identified as factors that influence system success over many decades. However, DeLone and McLean postulated that use and user satisfaction are reciprocally interdependent. Baroudi and his colleagues attempted to apply the theory of reasoned action to theoretically explain the causal relationship between user satisfaction and use (H_5). The theory of reasoned action suggests that a user's attitude (A) toward the behavior and subjective norm (SN) will determine the person's behavioral intention (BI). If user satisfaction is considered to be as an attitude, it will influence user's (voluntary) behavioral intention to use the system and actual use. This positive relationship between user satisfaction and use is empirically supported in a voluntary system and in a quasi-volitional system . Other studies such as Torkzadeh and Dwyer report a reciprocal relationship between the two (from user satisfaction to usage and from usage to user satisfaction.

In light of theoretical arguments and empirical evidence, two hypotheses are developed to test the reciprocal dependence. During a semester, students use an e-learning system and its outputs (information) first to generate their response to indicate the level of confidence and satisfaction. The level of satisfaction in turn is likely to influence the frequency of use and actual amount of time using e-learning systems. In contrast, if user satisfaction is viewed as a surrogate measure of e-learning systems success, then it makes theoretical sense to consider system usage as an antecedent to student user satisfaction. Therefore:

H_5. There will be a positive relationship between user satisfaction and use.

H_6. There will be a positive relationship between use and user satisfaction.

System use and individual impact

Individual impact is defined by DeLone and McLean as "the effects of information on the behavior of the recipient" In measuring IS success, a wide range of empirical measures are used as dependent variables including rate of performance, test scores, decision effectiveness (i.e., average time to make a decision, the confidence in the decision made), and changes in decision behavior.

In an e-learning context, individual impact can be interpreted as "learning outcomes." Empirical measures of e-learning outcomes includes perceived satisfaction with on-line courses, perceived satisfaction of delivery medium, perceived learning outcomes, and actual learning outcomes (course grade). Indicators of perceived learning outcomes include perceived quality of learning experience, the degree of improvement in the student ability to understand the basic concepts and to communicate about the subject, to identify the central issues, and to integrate facts and develop generalization.

In a voluntary system context, if the user avoids its use, information systems will be discontinued to result in the total failure with no individual impacts at all. Conversely, if the system is used continuously and frequently, it will result in higher and greater impact on a user's individual performance. This relationship between use and individual impact is empirically tested and well supported in the IS literature. Although there are seven empirical studies compiled by to support this relationship, there exists a paucity of empirical research that has investigated the relationship between e-learning systems use and e-learning outcomes due to the mandatory nature of e-learning system usage. We acknowledge the work of Freeze et al. who found a positive relationship between use and system success but as noted earlier, are argue that this model is mis-specified. Since we are attempting to assess the validity of the DeLone and McLean model in an e-learning context, in the current study we adapted the original hypothesis of the DeLone and McLean model.

H_7. There will be a positive relationship between systems use and e Learning Outcomes.

User satisfaction and individual impact

There is abundant prior research that examines the relationship between user satisfaction and individual impact . In addition, DeLone and McLean cite seven empirical studies that support a positive relationship between user satisfaction and individual impact. In an e-learning context, a study of Eom et al. examined the determinants of students' satisfaction and their perceived e-learning outcomes and found that user satisfaction is a significant predictor of learning outcomes. Therefore:

> H_8. There will be a positive relationship between user satisfaction and e-Learning Outcomes.

Research Methodology

The eight hypotheses were tested using the structural equation modeling (SEM) tool PLS graph version 3.0, build 1126. In the following section we first describe the development of the survey instrument and specific measures used to assess the variables are identified. Next, we justify the use of PLS as the method of data analysis. Scale reliability and validity of data are then reported. Finally, we present structural model results.

Survey Instrument

Wang, Wang, and Shee first explored whether IS success model of DeLone and McLean can be extended to investigate e-learning systems success in organizational contexts. Based on DeLone and McLean's updated IS success model, Wang et al developed and validated a multi-dimensional model for assessing e-learning systems success (ELSS) from the perspective of the employee (e-learner). The study conceptualized the construct of e-learning systems success, provided empirical validation of the construct and its underlying dimensionality, and developed a standardized instrument with desirable psychometric properties for measuring e-learning systems success.

Using the survey originally developed by Wang, Wang, and Shee for the business e-learning environment, the authors eliminated the questions addressing service quality, and slightly modified the language of some of the items of the ELSS to better fit the higher education environment and the perspective of the university online student. Modified items were reviewed by a group of online students, administrators, and faculty to ensure content validity and that the focus of each item was not changed from the ELSS instrument. The service quality questions were eliminated per the request of the online learning administrator who provided access to the university's online students and online testing resources. The survey instrument consisted of 36 items using a seven point Likert scale ranging from "strongly disagree" to "strongly agree." In addition, respondents were asked six demographic-type questions.

The population was undergraduate and graduate students that were enrolled in an online course at a large university located in the Midwest United States. We collected the e-mail addresses from the student data files achieved with every online course delivered through the online program of the university. Invitations to reply to the survey were administered online at the time of log-in to 2,156 unique students. Of those students invited, 809 students volunteered responses with 674 surveys being complete and usable for a response rate of 31.3%. Just under 20% of students were from the College of Business, 28.3% from the College of Education, 19.3% from the College of Health and Human Services, 8.6% from the College of Liberal Arts, and 11.4% from the College of Science, Technology and Agriculture. The remaining 7.4% of students represented two other colleges. Just over 5% of students did not declare their major.

To assess the extent of methods bias in our study the Harman one-factor test was performed following the approach described by Podsakoff et al. (1984) and Schriesheim (1979). This can best be described as a diagnostic technique for assessing the extent to which method bias may be a problem (Podsakoff, Mackenzie, Lee, & Podsakoff, 2003). According to this test, if a single factor emerges from the exploratory factor analysis or one factor accounts for more than fifty percent of the variance in the items, methods bias is present (Mattila & Enz, 2002). All of the items measuring the five constructs were entered into a common factor analysis with OBLIM rotation. The results revealed a five-factor structure with no one factor accounting for more than fifty percent of the variance. Therefore method bias, per se, cannot explain our study results.

Method Of Data Analysis

The use of PLS to test the research model (Figure 2) was justified on the following grounds. First, PLS is well suited to the early stages of theory building and testing as is the case with this research study. It is particularly applicable in research areas where theory is not as well developed as that demanded by LISREL. As pointed out by Seddon and Livari , the underlying theory of the DeLone-McLean model is not very strong. PLS is orientated towards prediction which is the major purpose of this paper. Second, PLS does not imply parametric assumptions of multivariate normal distribution . This is not the case with the maximum likelihood estimation method used in LISREL, which assumes a multivariate or normal distribution. In the present study, the data were not normally distributed (17 items showed Skewness > $\pm$1.0; 11 items showed Kurtosis > $\pm$2.0). For these two reasons, we believe that PLS analysis may produce more reliable statistical testing outcomes with the current dataset which violated normal distribution assumption.

Measurement (Outer) Model Results

As Figure 2 includes a reciprocal relationship between user satisfaction and system use that could not be tested at the same time in PLS, we tested two models. The first model (Model 1) showed user satisfaction to impact system use (H_5); the second model (Model 2) showed system use to impact user satisfaction (H_6). Due to the fact that the latent variable can be only indirectly measureable by using more than one observable variable, the issue of *validity* in SEM is an important matter. Validity is concerned with the extent to which each observed variable accurately defines the construct. Each measurement item on a survey instrument is assumed to reflect only one latent variable and each item is related to one construct better than to any others. This property (uni-dimensionality) of the construct must be confirmed. There are two elements of construct (factorial) validity in PLS: convergent validity and discriminant validity.

Convergent validity is defined as the extent to which indicators of a latent variable converge or share a high proportion of variance in common . It is established when all indicator (observed) variables load highly on their assigned factors, 0.5 or higher. Loadings of 0.5 are considered acceptable if there are additional indicators in the block for comparative purposes . Ideally however, they should be 0.7 or higher. Additionally, each of the measurement items should load with an acceptable *t*-value on its latent construct. The acceptable t-value is when p-value is at the .05 α level or less.

In a preliminary measurement model, two items in both models exhibited loadings below the 0.5 threshold (questions 13 and 14) and were subsequently dropped from further analysis. Table 1 shows the items belonging to the constructs for the revised measurement models, and indicates that most of the loadings are above .9 and that both models (Models 1 and 2) exhibit very similar results. The smallest loading is .7520. The *t*-values for the measurement items on their respective latent constructs are also reported.

Construct reliability was assessed with two measures – the composite reliability measure of internal consistency and average variance extracted (AVE). The internal consistency measure is similar to Cronbach's alpha except the latter presumes, a priori, that each indicator of a construct contributes equally (i.e., the loadings are set to unity). The internal consistency measure, which is unaffected by scale length, is more general than Cronbach's alpha, but the interpretation of the values obtained is similar and the guidelines offered by Nunnally and Bernstein can be adopted. Table 1 shows that all reliability measures for both models were above the recommended level of 0.70, thus indicating adequate internal consistency. The average variance extracted scores (AVE) were also above the minimum threshold of 0.5 and ranged from 0.728 to 0.923. The formula of AVE is: $\sum\lambda_i^2 /n$ where λ_i is the factor loading of each observed variable on its corresponding construct and n is the number of observed variables on each construct.

Discriminant validity is established when variables do not cross-load on two or more constructs. In other words, each observed variable loads highly on its theoretically assigned construct and not highly on other constructs. In PLS, discriminant validity was assessed using two methods. First, item loadings and the cross-loadings of the constructs and the measures were examined. This analysis revealed that the correlations of each construct with its measures were higher than the correlations with any other measures.

Second, the square root of the average variance extracted (AVE) for each construct was compared with the correlation between the construct and other constructs in the model. Adequate discriminant validity is demonstrated when the square root of the AVE for each construct is larger than the correlation between the construct and any other construct in the model. Table 2 shows that the square root of each AVE is larger than any correlation among any pair of latent variables thus indicating discriminant validity.

Table 1: Measurement model results

		Model 1		Model 2	
Constructs and Associated Variables	**Measurement Items**	**Loading**	***t*-tatistic[#]**	**Loading**	***t*-statistic[#]**
System Quality, IC = 0.930 AVE = 0.728					

Q1	The system is always available.	**0.7520**	**26.6390**	**0.7520**	**23.6528**
Q2	The system is user friendly.	**0.8992**	**41.7976**	**0.8992**	**41.1767**
Q3	The system provides interaction between users and the system.	**0.9013**	**37.8572**	**0.9013**	**38.7567**
Q4	The system has attractive feature that appeal to the user.	**0.8736**	**33.0827**	**0.8736**	**34.1619**
Q5	The system provides high-speed information access.	**0.8297**	**30.3190**	**0.8297**	**29.8376**
Information Quality, IC = 0.955, AVE = 0.811					
Q6	The system provides info. that that is exactly what you need.	**0.9075**	**3.6700**	**0.9075**	**3.9493**
Q7	The system provides info. that is relevant to learning.	**0.9320**	**4.7703**	**0.9319**	**4.8429**
Q8	The system provides sufficient information.	**0.9015**	**2.2393**	**0.9016**	**2.2247**
Q9	The system provides information that is easy to understand.	**0.8715**	**1.8835**	**0.8715**	**1.9052**
Q10	The system provides up-to-date information.	**0.8889**	**4.8095**	**0.8889**	**4.8640**
System Use IC = 0.926 AVE = 0.862					
Q11	I frequently use the system.	**0.9362**	**31.8443**	**0.9364**	**30.7917**
Q12	I depend upon the system.	**0.9204**	**31.7533**	**0.9202**	**30.4530**
User Satisfaction IC = 0.960 AVE = 0.923					
Q15	I think the system is very helpful.	**0.9592**	**119.2642**	**0.9585**	**126.4927**
Q16	Overall, I am satisfied with the system.	**0.9625**	**76.1124**	**0.9632**	**75.7108**
Individual Impact IC = 0.964 AVE = 0.870					
Q17	The system has a positive impact on my learning.	**0.9320**	**70.9795**	**0.9320**	**69.7720**
Q18	Overall, the performance of the system is good.	**0.9516**	**58.3445**	**0.9516**	**60.4973**
Q19	Overall, the system is successful.	**0.9533**	**67.5844**	**0.9533**	**70.5743**

Q20	The system is an important and valuable aid to me in the performance of my class work.	**0.8924**	**58.4765**	**0.8924**	**59.7911**

Note: IC is internal consistency measure; AVE is average variance extracted. # All significant p <.05.

Table 2: Correlation among construct scores (square root of AVE in the diagonal)

	System Quality	Information Quality	System Use	User Satisfaction	Individual Impact
System Quality	**0.853**				
Information Quality	0.846	**0.901**			
System Use	0.492	0.513	**0.928**		
User Satisfaction	0.813	0.824	0.518	**0.960**	
Individual Impact	0.815	0.841	0.560	0.914	**0.932**

Structural (Inner) Model Results

Since PLS makes no distributional assumptions in its parameter estimation procedure, traditional parameter-based techniques for significance testing and model evaluation are considered to be inappropriate. LISREL and other covariance structure analysis modelling approaches involve parameter estimation procedures, which seek to reproduce as closely as possible the observed covariance matrix. In contrast, PLS has its primary objective the minimization of error (or equivalently the maximization of variance explained) in all endogenous constructs.

Consistent with the distribution free, predictive approach of PLS , the structural model was evaluated using the *R*-square for the dependent constructs, and the size, *t*-statistics and significance level of the structural path coefficients. The *t*-statistics were estimated using the bootstrap resampling procedure (100 resamples). The results for Model 1 are summarized in Table 3 and the results for Model 2 are summarized in Table 4.

The results show that both models explain a considerable portion of the variance in system use, user satisfaction and individual impact. First, Model 1 explains 29.5 percent of the variance in system use. This compares with 28.8 percent of the variance in Model 2. Second, 71.6 percent of the variance in user satisfaction is explained in Model 1; this compares to 73.2 percent in Model 2. Third, the R^2 values for the system outcomes construct in Models 1 and 2 are 82.7 and 84.6 respectively. The percentage of variance explained for these three primary dependent variables is greater than 10 percent implying satisfactory and substantive value and predictive power of the PLS model .

Table 3: Structural (inner) model results for model 1

	Path Coefficient	Observed t-value	Sig. Level
Effect on System Use (R^2 = 0.295)			
System Quality	+0.083	0.7098	ns
Information Quality	+0.212	3.1217	****
User Satisfaction	+0.280	3.1149	****
Effect on User Satisfaction (R^2 = 0.716)			
System Quality	+0.429	7.4355	****
Information Quality	+0.452	8.520	****
Effect on Individual Impact (R^2 = 0.827)			
System Use	+0.121	4.4330	****
User Satisfaction	+0.841	44.7382	****

p-values: **** <0.001, ns not significant

Table 4: Structural (inner) model results for model 2

	Path Coefficient	Observed t-value	Sig. Level
Effect on System Use (R^2 = 0.288)			
System Quality	+0.154	2.1176	**
Information Quality	+0.400	4.9603	****
Effect on User Satisfaction (R^2 = 0.732)			
System Quality	+0.393	7.2073	****
Information Quality	+0.445	7.8588	****
System Use	+0.089	3.1151	***
Effect on Individual Impact (R^2 = 0.846)			
System Use	+0.118	4.5737	****
User Satisfaction	+0.853	45.9517	****

p-values: **** <0.001, *** <0.010, ** <0.05

As can be seen from the results for Model 2 (this model assumes system use to directly influence user satisfaction), the two antecedent constructs (systems quality and information quality) hypothesized to affect system use are significant with information quality being the stronger predictor (β = 0.40, t = 4.96). These findings suggest that the quality of information produced by e-learning systems and the quality of the e-learning system affects the use of the system to deliver online courses. In contrast, in Model 1 (this model assumes user satisfaction to directly influence system use), the direct relationship between system quality and system use is not supported (β = 0.08, *t* = 0.71). There is however, support for a direct relationship between information quality and system use (β = 0.21, *t* = 3.12). H_1 and H_2 are thus supported in Model 2 but only H_2 is supported in Model 1.

System quality and information quality also have a significant direct impact on the perceived satisfaction of students who take web-based courses in both models with information quality being the stronger predictor (β = 0.45, *t* = 8.85 in Model 1 and β = 0.45, *t* = 7.85 in Model 2). H_3 and H_4 are therefore supported.

A comparison of Models 1 and 2 reveal some valuable insights into the reciprocally interdependent relationship between system use and user satisfaction (H_5 and H_6). In Model 1 user satisfaction is a significant predictor of system use (β = 0.28, *t* = 3.11) suggesting that the perceived satisfaction of students who take web-based courses affects the use of the system to deliver online courses. H_5 is thus supported. In Model 2, the results show that system use has a significant direct impact on user satisfaction (β = 0.09, *t* = 3.11) thus supporting H_6. However, we suggest that this finding be treated with caution. Even though it is statistically significant, the substantive value of the path must also be considered. Paths of 0.10 for example, at best, represent a one-percent explanation of variance. Meehl argues that anything lower may be due to what he has termed the crud factor where "everything correlates to some extent with everything else" (p. 204) because of "some complex unknown network of genetic and environmental factors" (p. 209). Therefore, at best, the direct influence of systems use on user satisfaction is weak even though it is statistically significant.

Finally, both system use and user satisfaction have a significant direct impact on system outcomes with user satisfaction being the significantly stronger predictor. H_7 and H_8 are therefore supported although the direct influence of system use on system outcomes is weak given the substantive value of the path (β = 0.12 in both models).

Although PLS estimation does not utilize formal indices to assess overall goodness-of-fit (GoF) such as GFI, CFI, chi-square values, NNFI and RMSEA, it can be demonstrated by strong factor loadings, high R^2 values and substantial and statistically significant structural paths . In accordance with the categorization of effect sizes by Cohen ; small: 0.02; medium: 0.13; large: 0.26), all of these effect sizes are large (R^2 values of between 0.29 and 0.85 in both models). Tenenhaus et al. have also developed an additional GoF measure for PLS based on taking the square root of the product of the variance extracted with all constructs with multiple indicators and the average R^2 value of the endogenous constructs. The GoF measure is .708 in Model 1 and 0.714 in Model 2. Both measures indicate very good fit .

In addition to the R^2 for the dependent constructs and the size, *t*-statistics and significance level of the structural path coefficients, the Stone-Geisser test of predictive relevance was performed to further assess model fit in PLS analysis. The *Q*-square statistic represents a measure of how well the observed values are reconstructed by the model and its parameter estimates. *Q*-squares greater than zero imply that the model has predictive relevance, whereas *Q*-squares less than zero suggest that the model lacks predictive relevance. *Q*-square is 0.06 for system use in Model 1 and 0.08 in Model 2, and 0.66 for user satisfaction and 0.73 for system outcomes in both models.

Conclusion and Discussion

This study empirically tested the DeLone and McLean model of information systems success model in a university e-learning context using structural equation modeling. The eight research hypotheses were tested by using a PLS analysis on the survey data. All of the hypotheses in this study received full support except one: the relationship between system quality and system use. Considering a dearth of empirical tests of the DeLone and McLean model, the contribution of this paper is that this is the first empirical test of the DeLone and McLean model of IS success in a university e-learning context, which is strictly involuntary use. This study is distinguished from an earlier study which tested an incomplete mis-specified DM model in university e-learning context, with no causal link between system use and user satisfaction.

This study used the appropriate structural equation modeling tool of PLS methodology, which does not assume normal distribution of data. As briefly discussed in a previous section, the data were not normally distributed (17 items showed Skewness > $\pm$1.0; 11 items showed Kurtosis > $\pm$2.0). Consequently, PLS is a more appropriate tool for analyzing skewed data than LISREL which uses the maximum likelihood estimation method, which assumes a normal distribution of data. The IS success model of DM is validated in quasi-voluntary use of a student information system in a US university, in mandatory use of a municipal financial and accounting information system in Oulu City Council of Finland, and in mandatory use of e-learning system in a United States university.

Our study in both model 1 and model 2 reached several useful conclusions.

- Perceived system quality and perceived information quality are very strong (high path coefficient) predictor of user satisfaction.
- Perceived user satisfaction is a very strong (high path coefficient) predictor of individual impact.
- Perceived system quality is an insignificant predictor of system use (model 1) or relatively weak predictor of system use (model 2).
- The direct influence of system use on user satisfaction is weak even though it is statistically significant.

In order for e-learning students to be successful, they must be provided with e-learning system that provides information they need and user-friendly. Although system quality (user-friendly system) has not directly contributed to predict individual impact, its impact is indirect. System quality and information quality have positive effects on user satisfaction. Information quality has also positive effects on system use, which in turn positively contributes user satisfaction. Therefore, all the antecedent variables are positively affecting e-learning outcomes either indirectly or directly. System quality and information quality are necessary conditions for e-learning success, but not sufficient conditions[4].

Table 5: Comparison of three empirical study results

		Voluntary	Mandatory		Mandatory	
		UIS Rai et al. LISREL	FIS Livari (M1) PLS	Livari (M2) PLS	E-LMS (This Study) (M1) PLS	(M2) PLS
FROM	TO	Path Coeff.	Path Coeff.	Path Coeff.	Path Coeff.	Path Coeff.
System quality	Use	0.09*	0.30*	0.45**	0.08 n.s.	0.15**
System quality	Satisfaction	0.30***	0.55***	0.48**	0.43****	0.39****
Info quality	Use	0.18***	-0.19n.s.	-0.12n.s.	0.21****	0.40****
Info quality	Satisfaction	0.52***	0.26*	0.28*	0.45****	0.45****
Use	Satisfaction			0.14*		0.09***
Satisfaction	Use	0.35***	0.28*		0.28****	
Use	Indi. Impact	0.55***	0.15n.s.	0.15n.s.	0.12****	0.12****
Satisfaction	Indi. Impact	0.33***	0.52***	0.52***	0.84****	0.85****

**** $p < .001$, *** $p < .010$, ** $p < .050$, * $p < .100$, n.s. = not significant

4 For example, course structures and course contents affect perceived learning outcomes Sean B. Eom, Nicholas Ashill, and H. Joseph Wen, "The Determinants of Students' Perceived Learning Outcome and Satisfaction in University Online Education: An Empirical Investigation," *Decision Sciences Journal of Innovative Education* 4, no. 2 (2006). The course contents, on the other hand, are also affected by e-learning technology, especially e-learning content management systems.

Table 5 compares the results of three empirical studies. The core of the IS success model of DM is the system use. In the case of voluntary use of the information system (see the results of Rai, et al.' study in Table 5), the DeLone and McLean model has a reasonable explanatory and predictive power. But in a mandatory use environment, this study suggested it has much more limited explanatory and predictive power. The yellow-colored areas in Table 5 illustrate that among those constructs except system use, the DeLone and McLean model has produced consistently positive relationships with no conflicts among the three studies (5 models). However, there are conflicting results between system use and all other constructs (uncolored areas) except between system use and satisfaction. The results of the study of Freeze et al. is excluded in Table 5, because its model is mis-specified with no specified relationship between system use and user satisfaction. This is a major shortcoming of their work. Many previous studies suggest that there is a relationship between both constructs and numerous studies have argued for a reciprocal relationship. The potentially important relationship between system use and user satisfaction is ignored in the Freeze et al. work. What is really interesting is the significant relationship between system quality and system use in the Freeze et al. work but this relationship becomes non-significant when the relationship between user satisfaction and system use is also considered.

The nature of e-learning system use is fundamentally different from the environment in which the DeLone and McLean model is developed. E-learning management systems differ fundamentally from information systems that require voluntary use such as decision support systems (DSS). First, e-learning management systems are mandatory systems whereas DSS are voluntary systems. System use in the DeLone and McLean model was empirically measured by the use of decision support systems (see Table 3 of DeLone and McLean 1992). Second, the success measures of e-learning systems are user satisfaction and/or perceived/real learning outcomes whereas DSS success is measured by the decision time, the depth of analysis, the number of decision alternatives generated, the confidence and satisfaction of the decision makers with the process and outcomes, and quality and effectiveness of the decision process and outcome. Third, Critical success factors (CSF) of both systems, consequently are quite different. CSF of e-learning systems include, but not limited, students' motivation, self-efficacy, self-managing learning behavior, and learning styles. Instructors' teaching strategies, and meaningful feedback. Moreover, course structure/contents, interactions among three entities (students, the instructor, and students). On the other hand, CSF of DSS may include system use, user's task environment (decision context – unstructured or semi-structured), user-training and experience, intuition and judgment of the user. In summary, the use of decision support systems is mandatory and the pre-requisite for DSS success, whereas e-learning systems put more emphasis on human dimension, and design dimension and less emphasis on technological dimension (information systems). For these reasons, the DeLone and McLean model has a very limited explanatory power for explaining the role of e-learning systems on the eventual objectives and outcomes of e-learning. In closing, the DeLone and McLean model may only be useful to test the role of information systems on the use of e-learning systems[5].

Usefulness of the DeLone and McLean Model in Future Empirical E-Learning Research

In the early stage of e-learning empirical research, TAM developed in the IS area in the 1980s has emerged as a useful model for explaining e-learning system usage and satisfaction and studying e-Learning systems adoption and integration into academic program. The TAM theorizes that system use is the critical intervening construct, determined by perceived usefulness and perceived ease of use. The TAM model has been extended by many other researchers. The unified theory of acceptance and use of technology (UTAUT), an extension of the TAM model, postulates that perceived usefulness and ease of use determine an individual's intention to use a system, which in turn, determines actual system use. The theory posits that the four key constructs directly determine usage intention and behavior . Moreover, gender, age, experience, and voluntariness of use are posited to mediate the impact of the four key constructs on usage intention and behavior. The DeLone and McLean model is a variation of TAM. The system use construct in both models is the crucial component. The TAM model focused the determinants of system use (perceived usefulness and perceived ease of use) whereas the DeLone and McLean model posited that perceived ease of use and information quality determine system use, which in turn, along with user satisfaction, determines perceived usefulness (individual impact in the DeLone and McLean model).

This study's findings move beyond the limitations of studies of online teaching and learning grounded in TAM by suggesting a model where

---

[5] When studying the impact of technology, technology components should be explored because they have differential effects. For further details, readers are referred to Milan Zeleny, "High Technology and Barriers to Innovation: From Globalization to Relocalization," *International Journal of Information Technology & Decision Making* 11, no. 2 (2012).

predictors of attitudes toward technology usage and actual usage predict course learning outcomes. One of the reasons it does so is that rather than treating user satisfaction and learning outcomes as separate dependent variables (i.e. Arbaugh, 2005), this model predicts and empirically supports a relationship between user satisfaction and learning outcomes. One potential implication of this relationship is the possibility for different types of learner knowledge to be needed for successful experiences with e-learning. Educational researchers recently have argued that in order to teach successfully in technology-mediated environments, instructors need to integrate knowledge of instructional approaches, technological platforms, and pedagogy in ways that do not exist when considering each independently. Our findings suggest that similar approaches also may be necessary for learners to successfully navigate these learning environments.

Another reason why these findings extend beyond those of TAM-driven studies is that it considers the role of content in shaping learners' attitudes and behaviors toward the e-learning system. Some recent studies have suggested that learner interaction with content is a significant factor in shaping learning outcomes in online management education settings , but this study extends that work by suggesting that content also may predict attitudes toward and usage of a system, which in turn predicts satisfaction with an e-learning system. This potential for the e-learning systems to mediate the relationship between learner-content interaction and learning outcomes should be examined in future research.

The importance of e-learning system use is primarily dependent on the information technology developmental stages and telecommunication infrastructure of each country. TAM and the DeLone and McLean model can be a useful model to facilitate the adoption of e-learning initiative in underdeveloped and developing countries. It is quite interesting to notice that in some countries such as South Africa the primary purpose of implementing e-learning systems is to supplement the face-to-face classroom activities, not to replace them. Therefore, understanding the factors that encourage or inhibit e-learning system usages by the instructor and the students is one of the important research topics in South Africa . In sum, TAM and the DeLone and McLean model can be useful tools in e-learning empirical research in developing countries. However, these models may not be no longer useful e-learning empirical research models in technologically advanced countries. Many of the earlier research investigated learner characteristics such as attitudes toward e-learning, prior experience with computers, and e-learning systems' characteristics such as ease of use, perceived usefulness, perceived creditability to identify factors affecting acceptance of asynchronous e-learning systems in voluntary environments. Even today, in addition to other factors such as institutional support, information technology infrastructure, e-learning acceptance model play an important role to investigate whether information technology infrastructure is the key determinants of e-Learning Systems Adoption and acceptance of e-learning initiatives in the countries with poor information technology infrastructure.

Empirical Studies of E-Learning Systems' User Satisfaction

The next wave of e-learning empirical research shifted its focus on the CSFs that determine user satisfaction. Eom and others found that all six factors—course structure, self-motivation, learning styles, instructor knowledge and facilitation, interaction, and instructor feedback—significantly influenced students' satisfaction. This is in accordance with the findings and conclusions in the literature on student satisfaction. E-learning systems comprised of a myriad of subsystems of human factors, design factors, and technological factors. Human factors include personality characteristics, learning styles, and instructor's attributes. Design factors include a wide range of constructs that affect effectiveness of e-learning systems such as learner control, learning model, course contents and structure, and interaction. The DeLone and McLean model we tested deals with use of information technology, information quality, and system quality as antecedents of perceived e-learning satisfaction. Thus, the DeLone and McLean model explains only part of CSFs of e-learning systems satisfaction.

Empirical Studies of E-Learning Systems Outcomes

Electronic delivery medium's effectiveness has been a long-standing issue and continues to persist. Nevertheless, some meta-analytic studies also suggest that e-learning system outcomes now equal and *in some cases* surpass those provided in classroom-based settings. Online education can be a superior mode of instruction if it is targeted to learners with specific learning styles (visual and read/write learning styles), learners with specific personality characteristics and with timely, helpful instructor feedback of various types. Although cognitive and diagnostics feedbacks are all important factors that improve perceived learning outcomes, meta-cognitive feedback can induce students to become self-regulated learners. Computer self-efficacy was positively linked to learning outcomes measured by the average test scores in e-learning. Johnson, Hornik and

Salas found that student self-efficacy and perceived usefulness of the system predicted perceived content value, satisfaction, and learning performance. Other system-related studies have examined attitudes and behaviors influencing course management system usage. Significant positive correlations were found among the three e-learning variables (self-efficacy, e-learner satisfaction, and perceived usefulness).

E-learning systems empirical research should shift the focus from the critical success factors of e-learning satisfaction to e-learning outcomes. This is one possible way of systematically organizing vast amounts of burgeoning empirical research. During the past decade, a large number of e-learning empirical studies were conducted to investigate the impacts of too numerous factors. At this stage of e-learning systems empirical research, it is necessary to find a unifying framework that can help us identify critical success factors of e-learning systems and weave these factors together to make their interrelationship unequivocal. For a knowledge-capturing perspective, focusing studies on predictors of e-learning outcomes could help accelerate the pace at which meta-analytical approaches can be applied to this research stream. This research presents a framework that links various past empirical research, from adoption of e-learning initiative, to critical success factors of e-learning systems satisfaction, and finally to critical success factors of e-learning outcomes that are comparable or superior than traditional face-to-face learning system[6].

References

Ahmed, Hassan M. Selim. "Hybrid E-Learning Acceptance Model: Learner Perceptions." *Decision Sciences Journal of Innovative Education* 8, no. 2 (July 2010): 313-46.

Alavi, Maryam, and Erich A. Joachimsthaler. "Revisiting Dss Implementation Research: A Meta-Analysis of the Literature and Suggestions for Researchers." *MIS Quarterly* 16, no. 1 (March 1992): 95-116.

Anonymous. "Worldwide Elearning Market Surges to $27.1 Billion in 2009." *Business Wire*, February 17 2010.

Arbaugh, J. B. "Is There an Optimal Design for on-Line Mba Courses?." *Academy of Management Learning & Education* 4, no. 2 (2005): 135-49.

[6] We thank for the editor-in-chief, professor Milan Zeleny, for his suggestions which improved the quality of this article.

Arbaugh, J. B., Michael R. Godfrey, Marianne Johnson, Birgit Leisen Pollack, Bruce Niendorf, and William Wresch. "Research in Online and Blended Learning in the Business Disciplines: Key Findings and Possible Future Directions." *Internet and Higher Education* 12 (2009): 71-87.

Baroudi, J. J., M.H. Olsen, and B. Ives. "An Empirical Study of the Impact of User Involvement on Systems Usage and Information Satisfaction." *Communications of the ACM* 29, no. 3 (March 1986): 232-38.

Barti, Henri, and Sid Huff. "Change, Attitude to Change, and Decision Support System Success." *Information & Management* 9, no. 5 (December 1985): 261-68.

Belardo, S., Kirk R. Karwan, and William Wallace. "An Investigation of System Design Considerations for Emergency Management Decision Support." *IEEE Transactions on Systems, Man, and Cybernetics* SMC-14, no. 6 (November-December 1984): 795-804.

Bernstein, Jum C. Nunnally, Ira H. *Psychometric Theory*. 3rd ed. New York: McGraw-Hill, 1994.

Butler, Deborah L., and Philip H. Winne. "Feedback and Self-Regulated Learning: A Theoretical Synthesis." *Review of Educational Research* 65, no. 3 (1995): 245-81.

Chin, Wynne W. "Issues and Opinion on Structural Equation Modeling." *MIS Quarterly* 22, no. 1 (March 1998): VII-XVI.

—. "The Partial Least Squares Approach to Structural Equation Modeling." In *Modern Methods for Business Research*, edited by G. A. Marcoulides, 295-336. Mahwah, NJ: Lawrence Erlbaum Associates, 1998.

Cohen, Jacob. *Statistical Power Analysis for the Behavioral Sciences* 2nd ed. Hillsdale, NJ: Lawrence Erlbaum Associates, publishers, 1988.

Colquitt, Jason A, Jeffry A LePine, and Raymond A. Noe. "Toward an Integrative Theory of Training Motivation: A Meta-Analytic Path Analysis of 20 Years of Research." *Journal of Applied Psychology* 85, no. 5 (October 2000): 678-707.

Davis, Fred D. "Perceived Usefulness, Perceived Ease of Use, and User Acceptance of Information Technology." *MIS Quarterly* 13, no. 3 (September 1989): 319-40.

Davis, Robert, and Don Wong. "Conceptualizing and Measuring the Optimal Experience of the Elearning Environment." *Decision Sciences Journal of Innovative Education* 5, no. 1 (January 2007): 97-126.

DeLone, William H., and E. R. McLean. "The Delone and Mclean Model of Information Systems Success: A Ten-Year Update." *Journal of Management Information Systems* 19, no. 4 (Sping 2003): 9-30.

DeLone, William H., and Ephraim R. McLean. "Information System Success: The Quest for the Dependent Variable." *Information Systems Research* 3, no. 1 (March 1992): 60-95.

Dick, W., and L. Carey. *The Systematic Design of Instruction*. 3rd ed. New York, NY: Harper Collins Publishers, 1990.

Doll, William J., and Gholamreza Torkzadeh. "The Measurement of End User Computing Satisfaction." *MIS Quarterly* 12, no. 2 (June 1988): 259-74.

Drago, William A., and Richard J. Wagner. "Vark Preferred Learning Styles and Online Education." *Management Research News* 27, no. 7 (2004): 1-13.

Eom, Sean B. *The Development of Decision Support Systems Research: A Bibliometrical Approach*. Lewiston, NY: The Edwin Mellen Press, 2007.

Eom, Sean B., Nicholas Ashill, and H. Joseph Wen. "The Determinants of Students' Perceived Learning Outcome and Satisfaction in University Online Education: An Empirical Investigation." *Decision Sciences Journal of Innovative Education* 4, no. 2 (2006): 215-36.

Falk, R. F., and N.B. Miller. *A Primer for Soft Modeling*. Akron, OH: The University of Akron Press, 1992.

Fishbein, Martin, and Icek Azjen. *Belief, Attitude, Intention and Behaviour: An Introduction to Theory and Research*. Addison-Wesley Series in Social Psychology. Reading, MA: Addison-Wesley, 1975.

Fornell, Claes R., and F. L. Bookstein. "A Comparative Analysis of Two Structural Equation Models: Lisrel and Pls Applied to Market Data." In *A Second Generation of Multivariate Analysis*, edited by C. Fornell, 289-394 New York: Praeger, 1982.

Fornell, Claes R., and David F. Larcker. "Evaluating Structural Equation Models with Unobservable Variables and Measurement Error." *Journal of Marketing Research* 18, no. 1 (1981): 39-50.

Fornell, Claes R., and David F. Larcker. "Structural Equation Models with Unobservable Variables and Measurement Error: Algebra and Statistics." *Journal of Marketing Research* 18, no. 3 (February 1981): 39-50.

Freeze, Ronald D., Khaled A. Alshare, Peggy Lane, and H. Joseph Wen. "Is Success Model in E-Learning Context Based on Student Perception." *Journal of Information Systems Education* 21, no. 2 (2010): 173-84.

Gefen, David, and Detmar Straub. "A Practical Guide to Factorial Validity Using Pls-Graph: Tutorial and Annotated Example." *Communications of the Association for Information Systems* 16 (2005): 91-109.

Geisser, S. "The Predictive Sample Reuse Method with Applications." *Journal of the American Statistical Association* 70 (1975): 320-28.

Gist, M. E., C.K. Stevens, and A.G. Bavetta. "Effects of Self-Efficacy and Post-Training Intervention on the Acquisition and Maintenance of Complex Interpersonal Skills." *Personal Psychology* 44, no. 4 (1991): 837-61.

Gist, Marilyn E., Catherine Schwoerer, and Benson Rosen. "Effects of Alternative Training Methods on Self-Efficacy and Performance in Computer Software Training." *Journal of Applied Psychology* 74, no. 6 (December 1989): 884-91.

Goslar, Martin D. "Capability Criteria for Marketing Decision Support Systems." *Journal of Management Information Systems* 3, no. 1 (Summer 1986): 81-95.

Hair, J.F., W.C. Black, Barry J. Babin, and Rolph E. Anderson. *Multivariate Data Analysis*. 7th ed. Upper Saddle River, New Jersey: Prentice Hall, 2010.

Hayashi, Albert, Charlie Chen, Terry Ryan, and Jiinpo Wu. "The Role of Social Presence and Moderating Role of Computer Self Efficacy in Predicting the Continuance Usage of E-Learning Systems." *Journal of Information Systems Education* 15, no. 2 (Summer 2004): 139-54.

Holsapple, C. W., and A. Lee-Post. "Defining, Assessing, and Promoting E-Learning Success: An Information Systems Perspective." *Decision Sciences Journal of Innovative Education* 4, no. 1 (January 2006): 67-85.

Hornik, S., and A. Tupchiy. "Culture's Impact on Technology-Mediated Learning: The Role of Horizontal and Vertical Individualism and Collectivism." *Journal of Global Information Management* 14, no. 4 (2006): 31-56.

Johnson, Richard D., Steven Hornik, and Eduardo Salas. "An Empirical Examination of Factors Contributing to the Creation of Successful E-Learning Environments." *International Journal of Human–Computer Studies* 66, no. 5 (May 2008): 356-69.

Kellogg, Deborah L., and Marlene A. Smith. "Student-to-Student Interaction Revisited: A Case Study of Working Adult Business Students in Online Course." *Decision Sciences Journal of Innovative Education* 7, no. 2 (July 2009): 433-54.

Landry, B. J. L., R. Griffeth, and S. Hartman. "Measuring Student Perceptions of Blackboard Using the Technology Acceptance Model." *Decision Sciences Journal of Innovative Education* 4, no. 1 (January 2006): 87-99.

Leidner, D. E., and Sirkka L. Jarvenpaa. “The Use of Information Technology to Enhance Management School Education: A Theoretical View.” *MIS Quarterly* 19, no. 3 (1995): 265-91.

Liaw, Shu-Sheng, Hsiu-Mei Huang, and Gwo-Dong Chen. “Surveying Instructor and Learner Attitudes toward E-Learning.” *Computers & Education* 49, no. 4 (2007): 1066-80.

Livari, Juhani. “An Empirical Test of the Delone-Mclean Model of Information System Success.” *The DATA BASE for Advances in Information Systems* 36, no. 2 (Spring 2005): 8-27.

Martins, L. L., and F. W. Kellermans. “A Model of Business School Students’ Acceptance of a Web-Based Course Management System.” *Academy of Management Learning & Education* 3 (2004): 7-26.

Means, B., Y. Toyama, R. Murphy, M. Bakia, and K. Jones. “Evaluation of Evidence-Based Practices in Online Learning: A Meta-Analysis and Review of Online Learning Studies.” Washington, DC: U.S. Department of Education, 2009.

Meehl, Paul E. “Why Summaries of Research on Psychological Theories Are Often Uninterpretable.” *Psychological Reports, Monograph Supplement 1-V66* 66 (1990): 195-244.

Mishra, P., and M. J. Koehler. “Technological Pedagogical Content Knowledge: A Framework for Teacher Knowledge.” *Teachers College Record* 108 (2006): 1017-54.

Mlitwa, Nhlanhla W.B. *Integration of E-Learning Systems into Academic Programmes in Modern Universities: A South African Perspectives.* Cape Town, South Africa: TVK e-NNOVATIONS, 2011.

Moore, M.G. “Editorial: Distance Education Theory.” *The American Journal of Distance Education* 5, no. 3 (1991): 1-6.

—. “Three Types of Interaction.” *The American Journal of Distance Education* 3, no. 2 (1989): 1-6.

Morgan, G., and J. Adams. “Pedagogy First: Making Web Technologies Work for Soft Skills Development in Leadership and Management Education”. *Journal of Interactive Learning Research* 20 (2009): 129-55.

Nemanich, Louise, Michael Banks, and Dusya Vera. “Enhancing Knowledge Transfer in Classroom Versus Online Settings: The Interplay among Instructor, Student, Content, and Context.” *Decision Sciences Journal of Innovative Education* 7, no. 1 (January 2009): 123-48.

Northrup, Pamela T. “Online Learners’ Preferences for Interaction.” *The Quarterly Review of Distance Education* 3, no. 2 (2002): 219-26.

Oliver, R., and A. Omari. "Student Responses to Collaborating and Learning in a Web-Based Environment." *Journal of Computer Assisted Learning* 17, no. 1 (2001): 34-47.

Olson, Margrethe H., and Blake Ives. "Chargeback Systems and User Involvement in Information Systems." *MIS Quarterly* 6, no. 2 (June 1982): 47-60.

Olson, Margrethe H., and Blake Ives. "User Involvement in Systems Design: An Empirical Test of Alternative Approaches." *Information & Management* 4, no. 4 (September 1981): 183-95.

Ong, Chorng-Shyong, Jung-Yu Lai, and Yi-Shun Wang. "Factors Affecting Engineers' Acceptance of Asynchronous E-Learning Systems in High-Tech Companies." *Information & Management* 41, no. 6 (2004): 795-804.

Peltier, James W., William Drago, and John A. Schibrowsky. "Virtual Communities and the Assessment of Online Marketing Education." *Journal of Marketing Education* 25, no. 3 (December 2003): 260-76.

Pintrich, P. R., D..A. Smith, T. Garcia, and W. J. McKeachie. *A Manual for the Use of the Motivated Strategies for Learning Questionnaire (Mslq)*. Ann Arbor: University of Michigan: National Center for Research to Improve Postsecondary Teaching and Learning, 1991.

Rai, Arun, Sandra S. Lang, and Robert B. Welker. "Assessing the Validity of Is Success Models: An Empirical Test and Theoretical Analysis." *Information Systems Research* 13, no. 1 (March 2002): 50-69.

Redpath, L. "Confronting the Bias against Online Management Education." *Academy of Management Learning & Education* 11 (2012): 125-140.

Saade, R. G. "Web-Based Educational Information System for Enhanced Learning, Eisel: Student Assessment." *Journal of Information Technology Education* 2 (2003): 267-77.

Saade, R. G., W. Tan, and F. Nebebe. "Impact of Motivation on Intentions in Online Learning: Canada Vs. China." *Issues in Informing Science and Information Technology* 5 (2008): 137-47.

Sarker, S., and J. Nicholson. "Exploring the Myths About Online Education in Information Systems." *Informing Science* 8 (2005): 55-73.

Schniederjans, Marc J., and Eyong B. Kim. "Relationship of Student Undergraduate Achievement and Personality Characteristics in a Total Web-Based Environment: An Empirical Study." *Decision Sciences: Journal of Innovative Education* 3, no. 2 (fall 2005): 205-21.

Seddon, Peter B. "A Respecification and Extension of the Delone and Mclean Model of Is Success." *Information Systems Research* 8, no. 3 (September 1997): 240-53.

Seddon, Peter B., and M-Y. Kiew. "A Partial Test and Development of the Delone and Mclean Model of Is Success." In *Proceedings of the International Conference on Information Systems*, edited by J.I. DeGross, S.L. Huff and M.C. Munro, 99-100. Atlanta, GA: Associations for Information Systems, 1994.

Simmering, Marcia J., Clay Posey, and Gabriele Piccoli. "Computer Self-Efficacy and Motivation to Learn in a Self-Directed Online Course." *Decision Sciences Journal of Innovative Education* 7, no. 1 (January 2009): 99-121.

Sitzmann, T., K. Kraiger, D. Stewart, and R. Wisher. "The Comparative Effectiveness of Web-Based and Classroom Instruction: A Meta-Analysis." *Personnel Psychology* 59 (2006): 623-64.

Stone, M. "Cross-Validatory Choice and the Assessment of Statistical Prediction." *Journal of The Royal Statistical Society* 36, no. 2 (1974): 111-47.

Sun, Pei-Chen, Ray J. Tsai, Glenn Finger, Yueh-Yang Chen, and Dowming Yeh. "What Drives a Successful E-Learning? An Empirical Investigation of the Critical Factors Influencing Learner Satisfaction." *Computers & Education* 50, no. 4 (2008): 1183–202.

Swan, Karen. "Virtual Interaction: Design Factors Affecting Student Satisfaction and Perceived Learning in Asynchronous Online Courses." *Distance Education* 22, no. 2 (2001): 306-31.

Tenenhaus, Michel, Vincenzo E. Vinzi, Chatelin Yves-Marie, and Carlo Lauro. "Pls Path Modelling." *Computational Statistics and Data Analysis* 48, no. 1 (2005): 159-205.

Teng, James T. C., and Kenneth J. Calhoun. "Organizational Computing as a Facilitator of Operational and Managerial Decision Making: An Exploratory Study of Manager's Perceptions." *Decision Sciences* 27, no. 4 (Fall 1996): 673-710.

Torkzadeh, G., and D. J. Dwyer. "A Path Analytical Study of Determinants of Information System Usage." *Omega* 22, no. 4 (1994): 339-48.

Venkatesh, V., and F.D. Davis. "A Theoretical Extension of the Technology Acceptance Model: Four Longitudinal Field Studies." *Management Science* 46, no. 2 (February 2000): 186-204.

Venkatesh, V., and M.G. Morris. "Why Don't Men Ever Stop to Ask for Directions? Gender, Social Influence, and Their Role in Technology

Acceptance and Usage Behavior." *MIS Quarterly* 24, no. 1 (March 2000): 115-39.

Venkatesh, V., M.G. Morris, G.B. Davis, and F.D. Davis. "User Acceptance of Information Technology: Toward a Unified View." *MIS Quarterly* 27, no. 3 (September 2003): 425-78.

Wang, Yi-Shun, Hsiu-Yuan Wang, and Daniel Y. Shee. "Measuring E-Learning Systems Success in an Organizational Context: Scale Development and Validation." *Computers in Human Behavior* 23, no. 4 (July 2007): 1792–808.

Wold, H. "Systems Analysis by Partial Least Squares." In *Measuring the Unmeasurable*, edited by Peter Nijkamp, Heiga Leitner and Neil Wrigley, 221-52. Dordrecht: Martinus Nijhoff, 1985.

Womble, Joy. "E-Learning: The Relationship among Learner Satisfaction, Self-Efficacy, and Usefulness." *The Business Review* 10, no. 1 (Summer 2008): 182-88.

Zeleny, Milan. "High Technology and Barriers to Innovation: From Globalization to Relocalization." *International Journal of Information Technology & Decision Making* 11, no. 2 (2012): 441-56.

CONTRIBUTORS

Editors

Mohamed Ridda Laouar
Assistant Professor
Department of Computer Sciences
University of Tebessa, Algeria
ridda_laouar@yahoo.fr

Sean Eom
Professor of MIS
Department of Accounting
Southeast Missouri State University
Cape Girardeau, MO 63701 USA
sbeom@semo.edu

Chapter One

Ouail Abroun
PhD Student, National School of Applied Sciences
Computer Science, Operational Research and Applied Statistics Laboratory
Abdelmalek Essaadi University
93000 Tetouan, Morocco
abrounouail@gmail.com

Abderrahim Tahiri
Associate professor, National School of Applied Sciences
Computer Science, Operational Research and Applied Statistics Laboratory
Abdelmalek Essaadi University
93000 Tetouan, Morocco
abderahim.tahiri@gmail.com

Noura Aknin
Associate professor, Faculty of Science

Computer Science, Operational Research and Applied Statistics Laboratory
Abdelmalek Essaadi University
93000 Tetouan, Morocco
aknin@uae.ma

Kamal Eddine EL Kadiri
Associate professor, National School of Applied Sciences
Computer Science, Operational Research and Applied Statistics Laboratory
Abdelmalek Essaadi University
93000 Tetouan, Morocco
elkadiri@uae.ma

Chapter Two

Dalel Kanzari
Assistant Professor, Quantitative Method Department
Faculty of Management and Economic Sciences Mahdia, Monastir University, 5100
Tunisia
kndalel@yahoo.fr

Khaled Mellouli
Professor, Quantitative Method Department,
Institute of High Commercial Studies, Carthage University, 2082
Tunisia
Khaled.Mellouli@ihec.rnu.tn

Chapter Three

Olgerta Tona
Ph.D. student
Department of Informatics,
Lund University School of Economics and Management, Lund, Sweden
Olgerta.Tona@ics.lu.se

Sven A. Carlsson
Professor
Department of Informatics,

Lund University School of Economics and Management, Lund, Sweden
Sven.Carlsson@ics.lu.se

Sean Eom
Professor
Department of Accounting
Southeast Missouri State University
1 University Plaza
Cape Girardeau, MO 63701, USA
sbeom@semo.edu

Chapter Four

Mostefai Mohammed Amine
Associate Lecturer
Ecole Supérieure Informatique
BP 68M, 16270, Oued Smar, Algiers, Algerai
m_mostefai@esi.dz

Mohamed Ahmed-Nacer
Professor
USTHB (Université Sciences et Technologie Houari Boumediène)
BP32, El Alia 16111, Bab Ezzouar
Algiers, Algeria
anacer@cerist.dz

Chapter Five

Afef Selmi
Ph.D. student, Department of Computer Science
RIADI Laboratory
University of Tunis El Manar
1068 Tunis, Tunisia.
Selmi.afef@yahoo.fr

Mohamed Mohsen Gammoudi
Full Professor, Department of Computer Science
RIADI Laboratory
University of Manouba
2011 Tunis, Tunisia.
gammoudimomo@gmail.com

Farah Harrathi
Assistant Professor, Department of Computer Science
RIADI Laboratory
University of Manouba
2011 Tunis, Tunisia.
Harrathi.farah@gmail.com

Chapter Six

Zohra Ameur
Associate Professor
Department of Electronics, LAMPA Laboratory
Faculty of Electrical and Computer Engineering
Mouloud Mammeri University (UMMTO).
Bp 17 RP, 15000 Tizi-Ouzou, Algeria,
z_ameur@mail.ummto.dz
ameurzohra@yahoo.com

Mina Boulifa
Phd Student
Department of Electronics, LAMPA Laboratory
Faculty of Electrical and Computer Engineering Mouloud Mammeri University (UMMTO).
Bp 17 RP, 15000 Tizi-Ouzou, Algeria,
mboulifa2@yahoo.fr

Ramy Nekkache
PhD student
Department of Electronics, LAMPA Laboratory
Faculty of Electrical and Computer Engineering
Mouloud Mammeri University (UMMTO).
Bp 17 RP, 15000 Tizi-Ouzou, Algeria,
nramy2@yahoo.fr

Soltane Ameur
Professor
Department of Electronics
LAMPA Laboratory
Faculty of Electrical and Computer Engineering
Mouloud Mammeri University (UMMTO).
Bp 17 RP, 15000 Tizi-Ouzou, Algeria,
ameursoltane@yahoo.com

Chapter Seven

Youssef Nahraoui
Ph.D. Student, Department of Physics
Metrology and Information Processing Laboratory,
Ibn Zohr University
Faculty of science BP 8106 Hay Dakhla,
80000 Agadir, Morocco.
nahraouissef@gmail.com
nahraoui21y@yahoo.com

El Houcein Aassif
Professor, Department of Physics
Metrology and Information Processing Laboratory,
Ibn Zohr University
Faculty of Science BP 8106 Hay Dakhla,
80000 Agadir, Morocco.
aassifh@gmail.com

Gérard Maze
Professor Emeritus
University of Le Havre
Normandy University
LOMC CNRS UMR 6294
75 Bellot Street, CS 80 54., 76058 Le Havre, France
gerard.maze@univ-lehavre.fr

Chapter Eight

Abdeldjalil Gattal
PhD. Student,
LAMIS laboratory
Université de Tébessa,
2002 Tébessa, Algeria
Ecole nationale Supérieure d'Informatique (ESI)
Oued Smar, Algeria
ad.gattal@mail.univ-tebessa.dz

Youcef Chibani
Professor
Speech Communication and Signal Processing Laboratory
Faculty of Electronics and Computer Science

University of Science and Technology Houari Boumedienne (USTHB)
16000 Algiers, Algeria.
ychibani@usthb.dz

Chapter Nine

Gherghout Youcef
Assistant Lecturer, Department Of Computer Science
Faculty Of Engineer
UBMA University (Badji-Mokhtar University Annaba)
B.P.12, Sidi Amar ANNABA
Algeria
oucef_207@yahoo.fr

Tlili Yamina
Lecturer, Department of Computer Science
Faculty of engineer
UBMA University (Badji-Mokhtar University Annaba)
B.P.12, Sidi Amar ANNABA, Algeria
guiyam@yahoo.fr

Chapter Ten

Bekkouche Amina
Phd Student
Department of Computer Science
Abou Bekr Belkaid University of Tlemcen
Rue 565 N 20, Hai Ben Badis NEDROMA
Tlemcen, Algeria,13600
amina_13600@hotmail.com
ami_bekkouche@mail.univ-tlemcen.dz

Sidi Mohammed Benslimane
Associate Professor
Evolutionary Engineering and Distributed Information Systems Laboratory
Computer Science Department
Djillali Liabes University of Sidi Bel Abbes
B.P. 89 Sidi Bel Abbes, 22000, Algeria
sidimohamed.benslimane@gmail.com
benslimane@univ-sba.dz

Merzoug Mohamed
Phd Student
Department of Computer Science
Abou Bekr Belkaid University of Tlemcen
895 Oudjlida, Tlemcen, Algeria, 13000
Mohamed.merzoug@mail.univ-tlemcen.dz

Chapter Eleven

Fatima El Morabet,
PHD Student, Department of Physic
Information Systems and Telecommunications Laboratory
Abdelmalek Essaâdi University
93000 Tetouan, Morocco.
fatimalmorabet@gmail.com,

Sanae Chairi
PHD Student, Department of Physic
Information Systems and Telecommunications Laboratory
Abdelmalek Essaâdi University
93000 Tetouan, Morocco.
chairi.sanae@gmail.com

Ahmed El Moussaoui
Vice President, Abdelmalek Essaâdi University
93000 Tetouan, Morocco
elmoussaoui@uae.ma

Noura Aknin
Professor, Department of Physic
Information Systems and Telecommunications Laboratory
Abdelmalek Essaâdi University
93000 Tetouan, Morocco.
aknin@ieee.org

Chapter Twelve

Atif Farid Mohammad
PhD Candidate, Department of Computer Science
University of North Dakota

Grand Forks, ND 58202 USA
atif.mohammad@my.und.edu

Hamid Mcheick
Associate Professor
Mathematics and Computer Science Department
University of Quebec at Chicoutimi,
555 Boulevard de l'Université, Chicoutimi
Quebec, G7H-2B1 Canada
Hamid_Mcheick@uqac.ca

Emanuel S. Grant
Associate Professor
Department of Computer Science
University of North Dakota
Grand Forks, 58202 USA
grante@cs.und.edu

Chapter Thirteen

Abdelaziz Amara Korba
Ph.D. Student, Department of Computer Science
Networks and Systems Laboratory
University of Badji Mokhtar
23000 Annaba, Algeria.
amarakorba.abdelaziz@gmail.com

Mehdi Nafaa
Associate Professor, Department of Computer Science
Networks and Systems Laboratory
University of Badji Mokhtar
23000 Annaba, Algeria.
mehdi.nafaa@univ-annaba.org

Salim Ghanemi
Associate Professor, Department of Computer Science
University of Badji Mokhtar
23000 Annaba, Algeria.
ghanemisalim@yahoo.com

Chapter Fourteen

Youness Jouihri
PhD Student, Laboratoire d'Electronique et de Communications – LEC
EMI, Université Mohammed V-Agdal – UM5A
BP 765, avenue Ibn Sina Agdal,
Rabat, 10000, Morocco
Jouihri.Youness@gmail.com

Zouhair Guennoun
Professor, Laboratoire d'Electronique et de Communications – LEC
EMI, Université Mohammed V-Agdal – UM5A
BP 765, avenue Ibn Sina Agdal,
Rabat, 10000, Morocco
zouhair@emi.ac.ma

Chapter Fifteen

Sidi Mohamed O. Moulaye Abdellahi,
Ph.D. Student, Department of Computer Science and Mathematics
Laboratory of Computer Science
University of Cheikh Anta Diop
B.P. 5005 Dakar-Fann, Senegal
sidimed_makhtour@yahoo.fr

Mohamed T. Kimour
Professor, Department of Computer Science,
Laboratory of Research on Embedded Systems (LASE)
University of Badji Mokhtar,
23000 Annaba, Algeria
kimour@yahoo.com

Mbaye Sene
Maître de conférences (M.C)
Department of Computer Science and Mathematics
Laboratory of Computer Science
University of Cheikh Anta Diop
B.P. 5005 Dakar-Fann, Senegal
ngalagne@yahoo.com

Chapter Sixteen

Leila Zemmouchi-Ghomari
Lecturer, Department of IT Management
UMBB Universty (Université M'hamed Bougara, Boumerdès),
35000, Boumerdès, Algeria
l_ghomari@umbb.dz
l_zemmouchi@esi.dz

Abdessamed Réda Ghomari
Associate Professor
LMCS (Laboratoire de Méthodes de Conception de Systèmes)
E.S.I (Ecole nationale Supérieure d'Informatique) ex. INI
16309, Algiers, Algeria
a_ghomari@esi.dz

Chapter Seventeen

Amrita Chaturvedi
Ph.D. Student, Department of Computer Science and Engineering
Indian Institute of Technology
Kanpur – 208016, Uttar Pradesh, India.
amrita@iitk.ac.in
amritaiitk@gmail.com

T. V. Prabhakar
Professor, Department of Computer Science and Engineering
Indian Institute of Technology
Kanpur – 208016, Uttar Pradesh, India.
tvp@iitk.ac.in
tvp@cse.iitk.ac.in

Chapter Eighteen

Sean Eom
Professor
Department of Accounting
Southeast Missouri State University
Cape Girardeau, MO 63701, USA
sbeom@semo.edu

Nicholas J. Ashill
Professor
School of Business and Management
American University of Sharjah
PO Box 26666, Sharjah, UAE
nashill@aus.edu

J. B. (Ben) Arbaugh
Professor
College of Business
University of Wisconsin Oshkosh
800 Algoma Blvd.
Oshkosh, WI 54901 USA
arbaugh@uwosh.edu

James L. Stapleton
Associate Professor
Department of Management and Marketing
Southeast Missouri State University
Cape Girardeau, MO 63701 USA
jstapleton@semo.edu